SECOND THOUGHTS

SECOND THOUGHTS

Myths and Morals of U.S. Economic History

EDITED BY

Donald N. McCloskey

New York   Oxford
OXFORD UNIVERSITY PRESS
1993

Published for the Manhattan Institute by Oxford University Press.

Second Thoughts started as a gleam in the eye of William Hammett of the MANHATTAN INSTITUTE, and has been raised to maturity with the Institute's help by Larry Mone. The Institute financed the writing and arranged the publication, but otherwise left the professor-journalists here assembled to their own thoughts. Mark Riebling did a fine professional job of editing the manuscript.

Oxford University Press

Oxford   New York   Toronto
Delhi   Bombay   Calcutta   Madras   Karachi
Kuala Lumpur   Singapore   Hong Kong   Tokyo
Nairobi   Dar es Salaam   Cape Town
Melbourne   Auckland

and associated companies in
Berlin   Ibadan

Copyright © 1993 by Oxford University Press, Inc.

Published by Oxford University Press, Inc.,
200 Madison Avenue, New York, New York 10016

Oxford is a registered trademark of Oxford University Press

Library of Congress Cataloging-in-Publication Data
Second thoughts : myths and morals of U.S. economic history / edited
by Donald N. McCloskey.
p.   cm.   Includes bibliographical references and index.
ISBN 0-19-506633-2
1. United States—Economic conditions.   2. United States—Economic
policy.   I. McCloskey, Donald N.
HC103.S43   1993
330.973—dc20   92–9096

9 8 7 6 5 4 3

Printed in the United States of America
on acid-free paper

Contents

Contributors

Lee J. Alston
Department of Economics
University of Illinois

Terry L. Anderson
Department of Agricultural
  Economics and Economics
Montana State University

Benjamin Baack
Department of Economics
Ohio State University

Lance E. Davis
Department of Humanities and
  Social Sciences
California Institute of
  Technology

Barry Eichengreen
Department of Economics
University of California at
  Berkeley

Price V. Fishback
Department of Economics
University of Arizona

Robert Higgs
Department of Economics
Seattle University

Peter J. Hill
Department of Economics
Wheaton College

Elizabeth Hoffman
Dean of Liberal Arts and
  Sciences
Iowa State University

Jonathan Hughes
Department of Economics
Northwestern University

Robert A. Huttenback
Department of History
University of California at Santa
  Barbara

Gary Libecap
Department of Economics
University of Arizona

Donald N. McCloskey
Departments of Economics and
  History
University of Iowa

Robert A. Margo
Department of Economics
Vanderbilt University

Susan M. Phillips
Member, Board of Governors
Federal Reserve System

Edward Ray
Senior Vice Provost for
  Academic Affairs
Ohio State University

Hugh Rockoff
Department of Economics
Rutgers University

Nathan Rosenberg
Department of Economics
Stanford University

Elyce J. Rotella
Department of Economics
Indiana State University

Julian L. Simon
Department of Business
  Administration
University of Maryland

Rita James Simon
Department of Law and Public
  Affairs
American University

Richard Sylla
Department of Economics
Stern School of Business
New York University

Peter Temin
Department of Economics
Massachusetts Institute of
  Technology

Mark Thomas
Department of History
University of Virginia

Paul Uselding
Dean, College of Business
  Administration
University of Toledo

John Wallis
Department of Economics
University of Maryland

Gary M. Walton
President, Foundation for
  Teaching Economics
University of California, Davis

Jeffrey G. Williamson
Department of Economics
Harvard University

J. Richard Zecher
President and CEO, UBS Asset
  Management (New York) Inc.

SECOND THOUGHTS

Introduction:
Looking Forward into History

DONALD N. McCLOSKEY

This is a book that examines the past as a way of preparing for our future. It brings together a number of leading historians who show that commonly accepted wisdom about our economic past is often wrong, and therefore misleading. They persuade us that we will master the future—especially our economic future—only when we understand the lessons of our past. The quickest route to economic wisdom in our time, it turns out, is a detour through the nineteenth and early twentieth centuries.

Second Thoughts gives two dozen cases in point. Robert Higgs says: Forget what you think you know about the military–industrial complex; war is hell on the economy, too, and always has been. Julian and Rita Simon say: Fellow immigrants, stop worrying about the new immigrants; we have been through this before, and it worked out all right. Elyce Rotella says: Do not be misled by the sweet sound of "protective" legislation for women; the women's movement split in the 1920s over the issue, and may split again in the 1990s.

There is much to learn about the past of the American economy, its successes and its failures, the wise and the witless attempts to make things better. The Teapot Dome Scandal, argues Gary Libecap, originated as a wise attempt to solve the problem of drilling for oil in a common pool. "Free land" distributed by the Federal government in the nineteenth century, argue Terry Anderson and Peter Hill, tempted people to waste money chasing after a homestead.

Stories tell in economics as much as they do in literature. If we do not want today's financial scandals, like those of the 1930s, to lead to regulatory sclerosis, then the story by Susan Phillips and J. Richard Zecher about the birth of the Securities Exchange Commission (SEC) should go on the reading list. If we do not want today's big personal injury suits to lead to worse medical care and a more hazardous workplace, then we would do

3

well to read Price Fishback's story about workmen's compensation in coal mines eighty years ago.

The chief danger is that since the dramatic scientific progress of the early twentieth century, but especially since the 1960s, American policymakers have neglected the lessons of history. They do not need stories or metaphors; they have masters degrees in *Science,* these youngish individuals who disburse housing projects with a dash of sociology and foreign interventions with a dash of game theory. In the 1960s economists were a bit too sure of their 500-equation models and their ability to fine tune any economy around. Give us the computers, said the experts of Camelot, and we will finish the job.

The pride has led to tragedy, just as it did for Oedipus. It has led to the downfall of the powerful, on Pennsylvania Avenue in the 1960s and on Wall Street in the 1980s. A St. Louis housing project conceived by social scientists finally had to be blown up for landfill, a monstrous refutation of the doctrine that people after all are easy to predict and manipulate and social problems therefore easy to solve. Socialism has been the biggest failure in social engineering. The big American failure was Vietnam. A domestic political failure was Nixon and his colleagues. A financial one was Milken and his.

Economists finally have gotten the message. They have learned that they cannot predict in detail. For example, in 1969 an impatient regulator of the telecommunications industry decided to let MCI try an experiment between Chicago and St. Louis, thinking he could in this way stimulate Ma Bell to work harder. The unpredictable result, Peter Temin shows, was the breakup of AT&T. The expert economist in 1930 or in 1990 advises the government to manipulate confidence instead of changing reality. The unpredictable result, John Wallis shows, was the Great Depression of the early 1930s and a weak recovery in the early 1990s. Presidents Hoover and Roosevelt agree that technical intervention in agricultural markets— price supports, set asides, prohibitions on foreclosure—will save the American farmer from the disasters of the 1920s and 1930s. The unpredictable result, Lee Alston shows, was a bloated farm program taxing the poor to subsidize the rich. People, and especially politicians, cannot predict. If economists cannot predict in detail, then they cannot steer the economy with confidence.

Yet the future arrives tomorrow, and it would be unwise to meet tomorrow unprepared. What then is to be done? What is to be done is to get back to the humanism of history without sacrificing the real gains of economic science. What we can do if we cannot gain exact predictions is at least to gain approximate wisdom. The wise person owns an umbrella and carries it when the sky looks dark, even though he or she knows that weather prediction is poor at best. In a world desperate for insight into the future, the least we can try for is historical wisdom.

The craft of historical economics, a relatively new scientific field, is here applied to policy. The writers are all economists trained in the social engineering of the field. They are also historians. Economics has not always been a cumulative science, but since its official beginning in 1957 this historical economics, like labor economics or urban economics, has accumulated knowledge. A bibliography of what was accomplished down to 1980 contains 4,500 items; the findings since then have probably doubled. Historical economics has discovered Surprising Facts productive of wisdom. History is the best path to wise policy, though not much traveled.

Some of the wisdom is background. Behind talk of the former Third World lies a claim that imperialism is what made the West rich. Third World politicians still get votes from the claim. Jeffrey Williamson, then Lance Davis and Robert Huttenbach, show here that it is false. The Third World, says Williamson, has prospered since World War II, contrary to its impoverished image. Economic growth is never easy, and was no easier for Europe a century ago. Imperialism was no bar. On the other side the imperialism, gratifying though it was to the vanity of white and male Europeans, was no way to wealth. The largest empire the world has known, say Davis and Huttenbach, was a burden on the British people. India did not enrich Britain. Likewise the most ambitious foreign policy the world has ever known has not enriched the American people. NATO has been bad for business. The British, the French, the Americans, and recently our friends the Russians have learned in the end that an empire is a pain in the pocketbook.

So too, as it turns out, is Big Science. Behind talk of American educational failure lies a belief that the solution should somehow involve Big Science and America's preponderance in Nobel Prizes. Big Scientists have had their way since World War II, accelerated by the unexpected Soviet launching of *Sputnik,* with the result that America now has the best science in the world and the worst schools. Nathan Rosenberg shows that the Big Science claim is wrong, that science has commonly followed the economy rather than leading it. Good basic science, notwithstanding the claims of British astronomers and American biologists, does not make for a good life in the country paying for it. The science gets copied and developed by foreigners.

Such historical wisdom is coming back into favor with economists. Economic science today needs it. The new, lethal, and wisdom-producing criticism of the economist's predictions is that economics is a part of itself. An economics offering concrete predictions is trying to predict itself, like someone trying to lever oneself up with a crowbar. That is the technical reason historical economics is good for policy: economics cannot supply the social engineering it promises, so it had better learn the stories. The practical reasons are already implicit in American proverbs, collection points of wisdom.

One American proverb is, What you hit is more important than what you aim at. Free land for homesteaders, argue Terry Anderson and Peter Hill, was a noble experiment, but its very freeness caused the land to be badly used. Lawmakers and journalists talk about policy "designed" to do such and such, but seldom ask whether the "design" in fact works. The SEC, in the news today for its attacks on the rich and smart, was designed in the 1930s, as Phillips and Zecher argue, to help the poor, but ended up helping old wealth instead. The golden rule strikes again: those that have the gold, rule. Policies to "save jobs," as Elizabeth Hoffman finds in American history, do not always save jobs. One can see this in Poland, with valueless jobs in steel mills that belong in another country, and fewer jobs therefore in other parts of the Polish economy. One can also see it in America's nineteenth century. Hugh Rockoff describes the experience with price controls, the oldest of economic policies designed to do the impossible. Most rulers, from Hammurabi to Nixon, have given it a try.

History often repeats itself, but not always. We may be able to avoid the shooting war that grew out of economic competition between Britain and Germany in the late nineteenth century if we realize how similar is today's panic about Japanese competition. The Reagan Revolution was bound to fail. Paul Uselding points to a long history of confidence that government can know what the technological future will bring, and a long history of failure. Benjamin Baack and Edward Ray note the modern sound of the deal cut a century ago: give us the defense contracts, said the Northeast, and we will drop our opposition to the Federal income tax. Richard Sylla compares the deregulation of banking in the 1830s with that of the 1980s. What went wrong in the 1980s, unlike the 1830s, was that we deregulated by half measures. We got the Savings and Loan (S&L) Scandal as a result.

Of course, we need good myths, like those of the Founding Fathers and the Englishman's freedoms, but the bad myths get in the way of prudence. Certain historical myths are still abuilding, such as the one that Robert Margo exposes that the Civil Rights laws explain whatever success African-Americans have had; in fact, the people helped themselves, to education. Other myths have developed a history of their own, such as the myth Barry Eichengreen exposes that the period of the classic gold standard, 1879–1914, was golden all round. When Franklin Roosevelt finally decided to go off the gold standard his director of the budget said, "Well, this is the end of Western Civilization." It was not. Jonathan Hughes here exposes the myth of rugged individualism. Americans are individualists, all right, but since the Puritans they have exhibited a gift for collectivism as well, "precocity in the use of institutions of democratic coercion. . . . Big government was what we wanted and what we got." Americans have always believed that the government was a pork barrel.

Beyond the proverbs the rereadings of the American past illustrate the refined common sense called economics. A wiser because more historical economics is not antieconomic. On the contrary, Mark Thomas reviews the history of the trade deficit from Plymouth Rock to the Hard Rock Cafe, pointing out that the trade deficit is the same thing as the investment surplus: We either export goods or we export IOUs. The United States was built on an investment surplus, such as we now have, despite fears that foreigners are "buying up the country." Gary Walton tells the story of Robert Fulton's attempt to extract the whole profit from steamboating on the Mississippi. Fulton failed, and the opening of the trade to talent resulted by 1860 in rates one tenth what they had been in 1810. Ingenuity responds to supply and demand. The same economics is exhibited in Uselding's triad of stories. The government provided "patronage," like a Renaissance prince, but the technology responded best when the patronees were decentralized, numerous, and competing.

The essays commissioned here are an introduction to economic thinking, as painless as real thinking can be. Teapot Dome and Fulton's Steam engine, wildcat banking and Hoover's reputation, equal rights for women, and wartime wage and price controls can all be rethought wisely in the economist's way.

That wisdom comes with age makes it hard for entire countries to be wise. A country run by thirty-year olds is likely to have a short memory and unreasonable optimism about the possibilities of prediction—witness again the postwar Communist regime, or America in the 1960s. A country is as old as its memory. It was a learned joke of Renaissance scholars to point out that "the ancients" were actually the juveniles, because they lived in the first age of the world. Someone complained to T. S. Eliot about studying the past: "We know so much more than the ancients." Replied he: "Yes; and it is the ancients that we know."

I

INTERNATIONAL RELATIONS AND FOREIGN AFFAIRS

"No man is an Island, entire of it self," quoth John Donne, expressing a sentiment more European than American. Americans have had little patience with the rest of the world, and had especially little in the feverish century of continent-settling, 1789–1890. Our motto had often been that of the teenager or of Greta Garbo: "I want to be left alone."

Yet America has been part of the world, economically speaking, since its beginnings. We cannot be left alone. The newspapers speak these days of the global economy as though it were a novelty, but the first Europeans in New England were fishermen reaching across the globe in the fifteenth century to fill orders in Bristol and Glasgow. The indentured servants from whom descended about a third of the members of the Daughters of the American Revolution were part of the labor market of old England. A struggle between the British and the French gave us independence and a profitable trade with both sides. Americans have always been tied to the world economy by 100 tiny threads, Gulliver-style.

So the second thoughts begin with the oldest and the newest American self-description, as economic citizens of the world. We thus begin with Williamson on the prosperity of the neighborhood, Simon and Simon on immigration, Davis and Huttenback on the pomp of empire, and Higgs on the economic catastrophe of war.

1

How Tough Are Times
in the Third World?

JEFFREY G. WILLIAMSON

The Third World is full of economic basket cases, and growth was much faster in the currently advanced nations when they were undergoing their industrial revolutions. . . . Birth rates in the Third World are enormous by any previous historical standard, and this burden is likely to persist. . . . Third World cities are out of control, and overurbanization is currently taking place there compared with nineteenth-century experience. . . . Inequality has been an essential attribute of modern economic growth, for without it no developing country in the past could have saved enough to meet the investment requirements of industrialization. . . . Famines are clearly caused when large weather-induced food deficits hit economies already struggling with food shortages generated by population problems. . . . Colonialism caused dependency. . . .

These and many other perceptions of the Third World, which intelligent citizens carry in their heads, are just plain wrong. History offers no support for any of them. Most scholars believed these myths to be true thirty or forty years ago, but nobody had bothered to collect the evidence. Few of us had even much interest in Third World problems then, and thus no economic historian was looking at the past to help us understand the Third World future. Today, an impressive accumulation of historical knowledge makes it possible to reject these myths.

THE MYTH OF INDUSTRIAL GROWTH

Given the focus of the popular press on famine in Africa, the debt crisis in Latin America, President Aquino's problems in the Philippines, and internal strife in the Middle East, it may come as some surprise to learn that, by historical standards, economic growth in the Third World over the past three decades has been spectacular. The World Bank tells us that annual gross national product has grown, on average, three times faster than that of Western Europe during the nineteenth century. Even after adjusting for the fact that the Third World has a bigger population burden, the difference is pronounced: Income per head grew at about 1 percent per year among the nineteenth-century industrializers while it has grown at more than 3 percent per year in the Third World as a whole. Furthermore, these comparisons understate the magnitude of the Third World performance. We are comparing the successful industrializers in Europe (Belgium, Denmark, France, Germany, Italy, Holland, Norway, Sweden, and the United Kingdom) with *all* the Third World countries—the rapidly growing newly industrialized countries (NICs), slowly growing India, and the African disasters combined. If we instead compare the nine fastest growing Third World economies with the NICs who overtook Britain in the nineteenth century, namely, Germany and the United States, then Third World growth has been almost four times greater than it was in the industrializing nations of the nineteenth century. The so-called Gang of Four—Korea, Hong Kong, Taiwan, and Singapore—yield an even bigger differential. In short, we are living in an unusual period of very rapid economic growth among the Third World industrializers, far more rapid than that achieved by the industrializers in the nineteenth century. Indeed, Korea has done in two decades what England did in seven or eight.

There was certainly no such expectation forty years ago. It came as a surprise since there was no reason to expect that the Third World would do better than the nineteenth-century industrializers. Indeed, even *that* growth performance was unexpected since economists did not appreciate just how far many countries in the Third World had come since the late nineteenth century.

It might be objected that there are two favorable conditions facing the Third World that were absent in the nineteenth century. First, the followers rely heavily on capital inflows from the leaders. Second, the followers borrow advanced technologies from the leaders. This response, however, fails to appreciate the fact that the followers relied on foreign capital just as extensively in the nineteenth century, and that absorption of new technologies requires considerable skill and management.

How, then, do we reconcile this spectacular growth with the political rhetoric that stresses the widening gap between the North and the South, that is, between the industrialized countries and the Third World?

A simple example may help. Consider two countries, one with income per person of $100, and the other $1,000. Let the poorer of the two grow much faster, 10 percent per year, than the richer, 2 percent per year. The gap widens by $10! It follows that the gap in living standards between rich and poor countries will continue to grow in spite of the spectacular growth rates in the Third World. We will be well into the twenty-first century before those gaps even begin to collapse.

THE MYTH OF POPULATION EXPLOSION

Economic history had done nothing to overturn that central pillar of pessimistic Malthusian* thought, the principle of diminishing returns: The quality of life, at least measured in per capita income, is likely to grow faster if its rate of population growth is slower. Are the currently developing nations so different from the industrializers of the past?

Population growth rates are certainly far higher in the Third World today than they were in nineteenth century Europe, but that fact does not imply that *birth rates* are so much higher there. Indeed, according to the 1984 *World Development Report* and other well-known studies, crude birth rates in England in 1821 were 40.8 per thousand, which were not much lower than those in East Asia in 1950, 42.5 per thousand. Crude birth rates were higher in the United States in 1820, 55.2 per thousand, than they were anywhere in the Third World around 1950. It is simply not true that birth rates are enormous by any previous historical standard.

What, then, explains the more rapid rates of population growth? Most of the answer lies with *death* rates. Life expectancies are far longer in the Third World today than they were in England in the early nineteenth century, and longer life expectancies make for greater numbers. In fact, this is one of the great social achievements of the twentieth century. If we could place an economic value on greater longevity, economic growth in the Third World would look even more impressive. In this sense, population growth can hardly be viewed as a bad thing.

Nor is it true that birth rates are likely to remain high in the future. Demographers have recently discovered that birth rates have already started to decline in much of the Third World, first in Latin America and now in Asia. If these trends persist, they imply a far faster shift from large to small families than was true in the currently industrialized countries.

THE MYTH OF URBANIZATION

Growth rates of 6 percent per year are not uncommon to Third World cities. The United Nations forecasts some cities to reach remarkable size by the turn of this century—Mexico City at 31.6 million, Sao Paulo at 26 million, and Cairo, Jakarta, Seoul, and Karachi each exceeding 15 million. By the year 2000, twenty of the largest twenty-five cities will be in the Third World.

This urban explosion understandably generates fear among municipal planners, and the manifest poverty in the Third World cities suggests to

*Malthus, Thomas Robert (1776–1834), English economist and moral philosopher most famous for his contribution to population studies. Malthus argued that because populations increase *geometrically* while means of subsistence increase only *arithmetically,* standards of living will decline unless population growth is checked by famine, war, or disease.

some that these countries are overurbanizing. Imagine the lives of the urban unfortunates, however, were they forced to remain in even poorer villages. Immigrants to cities have always been pulled by favorable employment conditions in the town relative to the countryside. Karl Marx and his intellectual legacy aside, there is little evidence to support the view that workers were pushed into cities by a rural labor surplus in either 1790 England or 1990 India. To repeat, they were pulled there by better jobs.

Furthermore, the Third World is managing its urban growth with far greater skill and resource commitment than did, say, Manchester in the 1830s. Density and crowding are less problematic, and public health is more advanced. As a result, while mortality in the cities far exceeded mortality in the countryside in the nineteenth century, the opposite is true of Third World cities today. One of the reasons is that the Third World allocates a far larger share of their investment to urban housing and social overhead than did England in the early nineteenth century. During nineteenth-century industrial revolutions, cities were starved for social overhead. To put it simply, Third World governments are far more responsive to city investment requirements for public health, housing, water supplies, and other critical inputs to the quality of life than were the nineteenth-century industrializers. This is not to say that problems and poverty are absent in Third World cities, but rather that if we are looking for urban "failure," then we are more likely to find it in a nineteenth-century European or American city than in Third World cities today.

THE MYTH OF INEQUALITY

For at least two centuries, mainstream economists and government officials have been guided by the belief that national product cannot be raised while simultaneously giving the poor a larger share (i.e., that there must be some tradeoff between equity and growth). After all, does not redistribution to the poor cut into the surplus available for saving and accumulation?

Since only the top economic classes had a political voice in early nineteenth-century Britain, taxes and tariffs usually favored the rich at the expense of the poor, and the conventional trade-off view reigned supreme. From Adam Smith on, British economists eulogized saving and decried generous poor relief. Even such critics of capitalism as Marx and Hobson accepted the Smithian assumption that the rich save a lot more of their surplus than do the poor, so that to take away their income would choke off the supply of investable resources for the development effort. The trade-off thesis made its way well into the twentieth century: In the 1950s, Pakistani development plans stated that rising inequality might be unpleasant, but it was essential to finance early industrialization; in the 1980s, President Ronald Reagan's and Prime Minister Margaret Thatcher's advisers seemed to believe that low savings rates were the primary cause of productivity slowdown, and that overgenerous welfare programs accounted for both.

Although it seems undeniable that the rich save more than do the poor—being rich, they are able to have more of everything—American history does not offer very strong support for the view that there must be some trade-off between equity and growth. The savings rate rose markedly from the 1830s to the 1890s, thereby making it possible to finance the enormous investment requirements of industrialization and urbanization. Inequality rose as well, as is so typical of industrial revolutions. The rise in inequality, however, was a very small source of the rise in the savings rate, just as the revolutionary move toward equality after the 1920s had little to do with the decline in American saving rates in the mid-twentieth century. Nor has British historical evidence been kind to the trade-off thesis. While inequality seems to have risen in Britain after 1820, the share of saving in national product failed to rise at all.

The resolution of this apparent paradox is likely to have two parts. First, it was the rise in the rate of return to private investment during industrial revolutions that encouraged savers to save more, and it was technological advance that drove up the rate of return to private investment. Second, when we look at history we began to appreciate what economists have come to call the *isolation paradox:* It does not make much sense for me to invest in sewage removal on my block if none of my neighbors do so, but I would be happy to make that investment if my neighbors do the same. My neighbors and I are not likely to engage in the communal investment, however, without the intervention of an activist government that attempts a fair distribution of the tax burden. The rise in the investment rate in developing nations throughout the nineteenth century partially reflects a growing communal sensitivity to the isolation paradox, and the high public investment rates in the Third World partially reflect a full appreciation of the problem.

Of course, critics have often pointed out that the gap between rich and poor in much of the Third World is enormous (but remember, for every "Bad Latin nation" there is a "Virtuous Asian" one). We can see this gap by comparing poverty in the countryside with opulence in the cities, or a city's shanty dwellers with its favored bureaucrats and nouveau riche. Since we consciously or unconsciously compare the contemporary Third World with our own contemporary industrialized economies, our egalitarian sensitivities are outraged. The more relevant comparison, however, is with Europe and America a century ago, when early industrialization was in full stride. Seen in this light, inequality in Latin America today is no more extensive than it was in Britain and America in the 1880s. All of the currently industrialized countries underwent an egalitarian drift after World War I, and there is no reason to expect that Third World countries will not undergo the same leveling as they enter the twenty-first century.

THE MYTH OF FOOD SHORTAGES

Although many people believe that famines in the Third World are caused by recent population surges, mass starvation in Asia and Africa goes back a long way. The Indian economist Amartya Sen tells us, for instance, that the first Ethiopian famine was recorded in the ninth century; a century ago, perhaps one-third of Ethiopia's population was killed by a great famine, making the Ethiopian famine of 1972–1974 (in which 200,000 died, of a population of some 27 million) seem modest by comparison. Similar stories can be told for the Sahel, the part of Africa that borders on the Sahara, and in disaster-prone nations such as Bangladesh.

Despite historical precedent, however, recent famines in the Third World have been interpreted with Malthusian pessimism. In the early 1970s, for example, many observers pointed to increasing famine as evidence that the world was running out of resources. The policy implications were to pour more resources into birth control, slow the rate of industrialization, and search for alternative technologies that might reduce dependence on increasingly scarce fossil fuels and raw materials. The Third World, mimicking nineteenth-century industrial revolutions and population booms, was hardly enthusiastic about these policy conclusions. The Malthusian pessimists, however, had the ear of the popular press, and soon most "informed" Westerners believed that per capita food availability was not rising in the Third World, or, even worse, that it was falling. Famines from Ethiopia, Bangladesh, and the Sahel seemed to clinch the case.

The historical evidence turns out to be inconsistent with all of the inferences. Rather than increasing, famines have actually disappeared from much of the world. Europe had its last famine in the 1840s, during the failure of the Irish potato crop. The last great famine in India was in Bengal in 1943, and China has not had a famine since 1949. The intensity of famines in the Third World already showed signs of easing off in the late nineteenth century, especially in British India, an improvement that cannot be explained away by better luck with monsoons. There is simply no evidence to support the view that twentieth-century population has made Third World agriculture more vulnerable to weather-induced food deficits.

Furthermore, international organizations' projections of the Third World's "carrying capacity" (the ability to produce calories to support their growing populations) have too often neglected the true source of famine. In case after case of famine in the Third World, we have been led to believe that the central cause is "food availability deficit" (FAD) induced by drought, unfavorable monsoons, and other acts of God. In recent times, however, FADs have not been a primary cause of Third World Famine. During the Great Bengal Famine of 1943, for example, when perhaps as many as 3 million died, Sen tells us that food supply dropped by only 5 percent. The same can be said about most other mid-twentieth-century famines in the Third World.

The explanation for modern famines is not big FADs, but rather the *distribution* of the FAD, or what Sen calls the lack of entitlements to food. A small change in local food supplies can create an enormous rise in price. Those who suffer are the poor who no longer have a claim on now-scarce food. Landlords, peasants, and sharecroppers do not die in modern famines, but landless agricultural laborers and village artisans do—unless, of course, the state intervenes.

Since the eighteenth century, the state has encouraged the construction of railroads linking the interior to ports, thus making it possible for regions in food surplus to export to regions of food deficit. So it was that the British in India were able to mute the impact of local famines during the late nineteenth century. Even in Western Europe before the advent of the railroad, road and river transport links served to eliminate local famine. As the industrial revolution ensued, self-sufficient agriculture disappeared and was replaced by specialization in cash crops; an increasingly commercialized and capitalist agriculture created an agricultural proletariat, hiring out for wages, and an increasingly industrial and urban economy traded for food. As more individuals became economically distant from the direct production of food, a larger number of people became more vulnerable to FAD shocks. In addition, old institutional social insurance devices had disappeared with the destruction of village and kinship relationships. Modern state-supported social insurance schemes slowly took their place. Where they were slow to emerge, famine and food riots persisted (as in Great Britain in the early nineteenth century or Meiji, Japan, in the early twentieth century), but emerge they did.

The moral of history for the Third World is that it is inadequate social insurance that causes most modern famines, not weather-induced food deficits. In addition, it is the more egalitarian and open economy that is more likely to develop these social insurance schemes, or what have come to be called *safety nets*.

THE MYTH OF COLONIALISM

Most of us would like to find reason to blame our failures on others rather than on ourselves. The Third World would like to believe that its dependency can be traced to colonialism. While colonial policy certainly did little to foster industrialization in the Third World from the 1870s to World War II, the historical evidence does not suggest that colonialism was a central cause of this form of economic dependency.

Sir Arthur Lewis made this point more than a decade ago, and it is well worth repeating. The so-called international economic order was established during the four decades prior to World War I. As Sir Arthur pointed

out, the "rest" could have responded to the "West" in two ways: industrialization or trade. Since they did not industrialize, it must be asked,
Why not? Was it colonial imperialism? If so, how do we account for the
fact that independent countries like Thailand, China, Japan, and some
Latin American countries failed to industrialize until the twentieth century?
There must have been other forces at work. One such force is that productivity was too low in domestic agriculture. Low productivity agriculture
implied a small internal market for industrial products. Since most developing countries begin by producing manufactures for the home market,
lack of domestic demand stifled industrialization. Low productivity agriculture also implied a small surplus for accumulation, and world capital
markets were too underdeveloped to offer significant external supplies of
finance until well into the twentieth century. By contrast, Britain and the
United States underwent a long period of agricultural productivity improvement prior to their industrial revolutions. Thus, the Third World
opted for trade and exported products from their plantation and mining
enclaves. Once started, these trends were reinforced by colonial and domestic policies that favored those entrenched interests.

CONCLUSION

These, then, are some of the most prevalent myths that economic history
has exploded over the past three or four decades. It is important that we
expose them, since bad policy and poor performance is often defended by
a past that never was. The most important moral, however, is that the
Third World (especially outside preindustrial Africa) has accumulated a
far better growth record over the past half-century than did most of Europe
in the nineteenth century. While that record has been helped by World
Bank loans, technological transfer, and larger markets for export products
in rich countries abroad, it has been achieved primarily by better-informed
policy and more efficient institutions. As so many developing nations deal
with debt crises, population problems, polluted cities, inequality, and temporary economic setbacks, it is important to remember that the nineteenth-
century industrializing countries experienced the same growing pains. Despite these problems, the Third World has done very well by the standards
of history since Great Britain started the world industrial revolution two
centuries ago.

2

Do We Really Need
All These Immigrants?

JULIAN L. SIMON
and RITA JAMES SIMON

Immigration is a major issue for the 1990s. The Immigration Act of 1990 increased the flow of immigration and altered its composition. It also set up an oversight committee to observe the bill's effects, which foreshadows further action. Although the passions, arguments, and public opinion about immigration are basically the same as they have always been, the economics of immigration are better understood now than before. Whether this will matter, however, is another question.

A BRIEF HISTORY OF PUBLIC OPINION

The Founding Fathers were sure of the value of immigration. In fact, one of their major complaints against King George III was that he hindered the peopling of the colonies from abroad. According to the Declaration of Independence, His Majesty had

> endeavored to prevent the population of these States; for that purpose obstruct-
> ing the Laws for Naturalization of foreigners; refusing to pass others to encourage
> their migration better, and raising the conditions of new appropriation of Lands.

The architects of the American system perceived a link between liberal immigration policy and economic growth; as James Madison said at the Constitutional Convention in 1787: "That part of America which has encouraged the foreigner most has advanced the most rapidly in population, agriculture and the arts."

American public opinion, however, has always opposed immigration. In the first half of the nineteenth century, the Irish immigrants in New York

19

and Boston were seen as unassimilable degenerates. America was becoming "the sewer into which the pollutions of European jails are emptied," one editorial stated.

Very much in this spirit, the Democratic Party's 1884 platform stated its opposition to the "importation of foreign labor or the admission of servile races unfitted by habit, training, religion or kindred for absorption into the great body of our people or for the citizenship which our laws confer."

By 1896, this populist paranoia had become part of the agenda of Francis Walker, commissioner general of the Immigration Service and first president of the American Economic Association. According to Walker:

> The question today is protecting the American rate of wages, the American standard of living, and the quality of American citizenship from degradation through the tumultuous access of vast throngs of ignorant and brutalized peasantry from the countries of Eastern and Southern Europe. . . . The entrance into our political, social, and industrial life of such vast masses of peasantry, degraded below our utmost conceptions, is a matter which no intelligent patriot can look upon without the gravest apprehension and alarm.

In the 1920s, sociologist E. A. Ross warned in the *Atlantic Monthly* of immigration's social dangers to our culture:

> The immigrant seldom brings in his intellectual baggage anything of use to us— the admission into our electorate of backward men—men whose mental, moral, and physical standards are lower than our own—must inevitably retard our social progress and thrust us behind the more uniformly civilized nations of the world.

During the 1920s the *Saturday Evening Post,* in particular, directed fear and hatred at the "new immigrant" from southern and eastern Europe:

> More than a third of them cannot read and write; generally speaking they have been very difficult to assimilate. . . . They have been hot beds of dissent, unrest, sedition and anarchy. . . . Is not the enormous expense of maintaining asylums, institutions, hospitals, prisons, penitentiaries, and the like due in considerable measure to the foreign born, socially inadequate aliens?

Later in the 1920s came attacks in the *Post* against "The Mexican Invasion":

> We can search in vain throughout the countries of Europe for biological, economic, and social conditions fraught with a fraction of the danger inherent in the immigration of peons to the United States."

National opinion polls since the 1930s consistently show a majority of Americans against immigration. For example, in May 1938 Americans were asked: "What is your attitude toward allowing German, Austrian, and other political refugees to come to the United States?" Their responses were:

Encourage even if we have to raise immigration quotas..........................5%
Allow them to come, but do not raise quotas 18%
With conditions as they are, we should keep them out......................... 68%
Don't know...9%

In January 1939, a survey asked:

> If you were a member of the incoming Congress, would you vote yes or no on a bill to open the doors of the United States to a larger number of European refugees than are now admitted under our immigration quota?

Eighty-three percent of respondents to this question said they would vote *no*. Between 1946 and 1990 the following question appeared on national polls at least eight times: "Should immigration be kept at its present level, increased, or decreased?" In 1953, 13 percent of Americans polled said immigration should be increased, but the percentage favoring increasing immigration levels ranged from 5 to 9 percent in all the other years.

The attitude in each generation may be characterized as: "The immigrants who came in the past were good folk. But the ones coming now are scum."

SOME NUMBERS

The population of the thirteen states was about 4 million in 1776; the rest of what is the present United States was thinly populated. Two centuries later, the population is about 240 million. If there had been no immigration, our current population would be less than half as large as it now is, even if the rate of natural increase had been extraordinarily high throughout our history. Nor would our economy have had advanced as it did without such high rates of immigration. Lest one think that the experience of the United States is unique, we should notice two other countries that have had both extraordinary rates of immigration, and very high rates of economic advance at the same time. Hong Kong's population rose from a low of 700,000 at the end of World War II, and 1.7 million in 1947, to more than 5 million now, and of course its economic development has been breathtaking. West Germany took in unprecedented numbers of immigrants from East Germany and elsewhere at the end of World War II, and experienced the "German miracle" in the same period.

With these two cases plus that of the United States in hand, a causal connection suggests itself. Many ask, however, What about Japan? Well, what about it? True, Japan is extremely homogeneous, and it has a policy of not admitting immigrants; however, in the first few years after World

War II, when Japan was experiencing its own "miracle," it took in large numbers of overseas Japanese. Japan is hardly a counterexample, then.

THE MYTH OF THE HUDDLED MASSES

Emma Lazarus's poem at the base of the Statue of Liberty, as beautiful as it is, has fostered a false impression of immigrants. The poem speaks of "huddled masses" and "wretched refuse," but immigrants generally are neither tired, poor, "huddled masses," nor are they "wretched refuse." Rather, they are usually young and vigorous adults, well-educated and highly skilled, with excellent earnings potential.

Even when Ms Lazarus wrote her heartwarming poem a century ago, immigrants compared favorably with the native population in occupational skill. P. J. Hill, who analyzed immigrants' wages and occupations in censuses and other sources from 1840 to 1920, with special attention to the 1890s, concluded in his 1976 book, *The Impact of Immigration into the United States,* that "immigrants, instead of being an underpaid, exploited group, generally held an economic position that compared very favorably to that of the native-born members of the society." Bernard Bailyn's research on the colonial-period Registry of Emigrants from Great Britain, published in his 1986 book *Voyagers to the West,* reveals that immigrants had excellent economic characteristics, being mainly young "useful artisans" and farmers.

Today, too, the economic "quality" of immigrants is high compared with that of natives. Immigrants are mainly in their twenties and thirties, the ages of greatest physical and mental vigor, when people are flexible about job location and therefore help the economy adjust to changing conditions. As of the late 1970s, 32.5 percent of the U.S. population was in the 20–39 age bracket, 46.3 percent of the immigrant cohort was also in that prime bracket at that time. Among immigrants who are admitted on the basis of their occupations rather than their family connections, 61.6 percent were between the ages 20 and 39. In addition to being young and vigorous, the immigrants possess higher educations and professional capabilities in greater proportions than did the native labor force. In 1980, some 16 percent of employed natives were professional and technical workers. The corresponding figure for recent immigrants was 26 percent.

Another source of evidence is the experience of countries such as Canada and Australia, which also traditionally have been countries of immigration, and share with us an English-speaking culture. There is uncanny similarity in the pattern from country to country—the propensity of immigrants to start new businesses (higher than natives), labor-force participation (higher than for natives, especially among immigrant women), and the age structure of immigration (young adults).

Together with the patterns throughout U.S. history, these data give us a solid basis for predicting future patterns of immigration. The contemporary organizations that strive to reduce immigration—Federation for American Immigration Reform (FAIR), The Environmental Fund, and Zero Population Growth (ZPG)—say that the situation now is different from the past in a key respect: The United States is more a welfare state than it was a century ago. This is true, but they place the wrong interpretation on this historical shift.

The anti-immigrationists assume that a tax-and-transfer system implies that immigrants are a burden on the natives. Our research using a major Census Bureau survey, however, shows exactly the opposite. Immigrant families on average pay more in taxes than do native families, and use less in welfare services. This finding makes sense because the immigrants bring marketable skills and come when they are in the prime producing years.

Loss of jobs to immigrants has always been a bogey. The data, however, suggest that immigration has not caused additional native unemployment. Though the analysis for the long sweep of U.S. history is less than satisfactory statistically, the United States has had much more immigration than have other nations, and it has not had much more unemployment throughout the decades. Furthermore, in the period of great unemployment—the 1930s—net migration became negative because more foreign-born persons left the United States than came to these shores. Immigration buffers the employment situation, increasing the labor supply when more workers are needed and turning in the other direction when the demand for workers falls.

A spate of recent studies using a variety of methods—across regions, across cities, census data, INS immigration data, and so on—confirm that immigration does not create native unemployment. Assessing the benefits of immigration throughout history is difficult. Contrary to common economic thought, immigration does not yield the same sorts of benefits deriving from comparative advantage as does trade in goods. Because immigrants perform services that require their physical presence, and because they themselves receive the full value of those services, the benefits are mostly diffuse, and are mixed with the activities of natives.

The contributions of immigrants have been especially evident in some periods. The influx of Jewish and other refugee scientists and scholars in the 1930s was a shot in the arm for American culture and learning, as the amazing record of Nobel prizes to immigrants clearly shows. The boost in the number of native scientists who won Nobel prizes also surely derives partly from their having the immigrant scientists as colleagues. In addition, at the most fundamental level of national security, the development of the atomic bomb hinged on the participation of immigrants such as Enrico Fermi, John von Neumann, and Stan Ulam, among many others.

Contemporary newspaper stories continue this historical saga with the disproportionate numbers of Vietnamese and other Asian immigrant youths achieving distinctions such as success in the Westinghouse Science Talent Search and in the competition for scholarships at the Julliard School of Music.

Historians of immigrants remark about the great effort exerted by immigrants to establish themselves in the United States. The historians assert that waves of immigration renew the vitality of this country. Such phenomena, however, have not yet been established systematically, and therefore we shall say no more about them here.

THE POLITICS OF RACISM

If immigration offers so much good, and so little bad, why has the public been against it throughout our history? There is certainly always a considerable amount of economic ignorance operating, both, of the facts cited earlier, and also of the understanding of diffuse and long-run effects that Bastiat taught us about so well a century and a half ago in "The Seen and the Unseen." Henry Hazlitt retaught the lesson as "Economics in One Lesson": The art of economics consists in looking not merely at the immediate but at the longer effects of any act or policy, and consists in tracing the consequences of that policy not merely for one group but for all groups.

Economic ignorance, however, is not the entire explanation of opposition to immigration, and may not be even the half of it. Dislike of other races and religions, fear of foreigners, and a desire to maintain cultural homogeneity certainly are powerful agents. These forces are not obvious today because there are taboos and even laws against expressing these sentiments. Given the historical record both in the United States and elsewhere, however, it would be naïve in the extreme to mistake the absence of public utterance for the absence of operation. In Canada, as recently as just before World War II, Prime Minister Mackenzie King wanted to keep Canada "free from too great a mixture of foreign streams of blood." "Jews," he said, "would pollute Canada's bloodstream and undermine Canadian unity." More recently, British Prime Minister Margaret Thatcher has stated her desire to keep Great Britain from "being swamped by people of a different culture." There is little doubt that if such sentiments could be publicly expressed without repercussions in the United States, we would hear them here, too.

Still another reason to believe that unexpressed motives are at work are some of the preposterous charges against immigrants by persons who cannot but know better. In the mid–1980s, persons as eminent as Senator Alan Simpson, the main author of the congressional legislation against illegal immigration, and Attorney General Edwin Meese, who urged the administration to support the legislation, claimed that the flow of illegal immigration promotes terrorism and drug smuggling. Senator Simpson even endorsed the view of FAIR that said allowing illegal immigration is the equivalent of allowing slavery in the United States before the Civil War.

CONCLUSION

Many more hard-working and talented foreigners would like to move to the United States than the immigration laws permit. These persons could speed the economic advance of the United States and raise the standard of living of natives. If one is interested in the political capacity of the United States vis-à-vis the rest of the world, then these potential immigrants could improve the competitive position of the United States (as Mikhail Gorbachev stated on U.S. television with respect to migrants from the then Soviet Union to the United States). As has been the case throughout our history, economic ignorance and ethnic prejudice oppose the immigration that could bring these advantages.

3

Do Imperial Powers Get Rich Off Their Colonies?

LANCE E. DAVIS
and ROBERT A. HUTTENBACK

Just as average Americans are proud of their country's ability to intervene decisively in world trouble spots from the Panama Canal to the Persian Gulf, so, too, did the Briton in the street once share in the glory of the British Empire. During the late nineteenth and early twentieth centuries, even the poorest laborer in the "Blacklands" took vicarious pleasure in the fact that the sun never set on the Union Jack, and that so much of the globe was painted red. These English agreed with Prime Minister Benjamin Disraeli in viewing Britain as "a great country, an imperial country, a country where your sons, when they rise, rise to paramount positions, and obtain not merely the esteem of their countrymen, but the respect of the world."

This enthusiasm, however, was not shared by most of those in the British government, to whom the vast imperial territories were largely an expensive luxury and a burden. Marxist critiques notwithstanding, colonial adventurism was never a source of vast profit for the imperial governments. The people's representatives grudgingly accepted a responsibility that was very popular with their constituents, but empires cost a lot of money— and volunteers to share the expense were hard to find. This was the same lesson that the United States, weighed down by the responsibility of defending the "free world," would later learn. Indeed, a look at the history of the British Empire helps to explain why NATO eventually became such a burden to the United States.

REDCOATS AND REBELLION

It is perhaps ironic that the United States finds itself caught in the same financial dilemma that plagued the British in the years before 1776. In the first half of the eighteenth century, British military expenditures had averaged about 10 shillings per person per year. Between 1756 and 1763, however, when Britain faced the French and their allies on the North American continent, that figure more than doubled. To cover those extra costs, which were largely incurred in defense of the American colonists, British taxpayers faced the highest levies in the Western world. In each year between 1756 and 1776 British taxes amounted to more than £1 per person per year—or about 6 percent of the island nation's net national product. The colonists, on the other hand, paid essentially nothing, and felt righteous in their determination to see that the situation did not change.

As early as 1730, the Pennsylvania Assembly had refused to spend any money to stop privateers from interfering with the colony's trade with the West Indies and from raiding settlements on Delaware Bay. That august body felt that "it was incumbent upon the British government to afford protection by sea." Later, when the British began to press for some partial repayment of the expenditures incurred during the French and Indian Wars, Benjamin Franklin blamed the entire conflagration on an attempt by the British to protect the rights of their traders to an obscure fort in Ohio (he seems to have overlooked the fact that the traders were American colonists, and the fort was Pittsburgh). In time, the British did manage to recoup a tiny fraction of their expenditures, but the total amount was less than £1.5 million. On a per capita basis, that figure works out to less than one-tenth of the burden shouldered by the residents of Britain.

The experience of the French and Indian War, exacerbated by Indian eruptions like Pontiac's rebellion and the pressing need to defend the western lands taken from the French, convinced the British that a more sustained effort was needed to force the colonists to pay for what was, after all, their own defense. On May 5, 1763, the Board of Trade was asked by its government:

> [I]n what mode least burdensome and most palatable to the colonies can they contribute toward the support and additional defense that must attend their civil and military establishments in consequence of the newly acquired territories and forts therein.

The authorities were not attempting to gain full repayment, but merely £145,000, and they would have settled for three-fourths of that amount.

To put these figures in perspective, at the time of the Revolution, colonial taxes ranged from a high £0.07 per capita in Massachusetts to a low of £0.01 in New Jersey; the average was about £0.03. In Europe taxes varied from £1.03 in Britain to £0.05 pounds in Poland.

Since the colonists enjoyed one of the highest standards of living in the world, the British government could never fathom why the North Americans so obdurately refused to share the costs of their own defense. The British never demanded full payment, but even the modest imposts they imposed—the Stamp, Sugar, and Navigation duties, most of which were already in use in the mother country—led to war and the permanent separation of the colonies from the mother country.

THE BEST DEFENSE

The empire was clearly a costly business, but, if "salutary neglect" had produced an indifference to common needs, than perhaps a stricter policy would bring more acceptable results. The British government, therefore, moved to implant centralized, autocratic rule in the last major bastion of their empire—Canada. The fruit of that policy proved equally bitter— within a few years rebellions erupted in both Upper and Lower Canada. An attempt to realize a defense contribution from the thirteen largely autonomous colonies had resulted in disaster, but a structure based on stringent controls had proved no more viable. What could be done?

Happily, Lord Durham had yet another institutional alternative: "responsible government" for the colonies of white settlement. The theory was simplicity itself. The royal governor would not only be paid by the colony, but he would appoint as his chief minister the representative of the majority in a locally elected legislature. Thus, a colony with responsible government would be essentially self-governing, but, more important, it would be responsible for its own defense.

The new institution spread rapidly across the empire—from Canada, to the six Australian colonies, to New Zealand, and on to South Africa. In London the bureaucrats believed that the conception was sound, and they confidently awaited proof of its success. It was successful, but not in the way they had imagined. In theory the self-governing colonies were to pay for their own defense and for hostilities within their domain, and, while empire-wide defense was to remain a British responsibility, those colonies were expected to make a substantial contribution.

On the basis of the American colonial experience, the results should, perhaps, have been predicted. Not only did the British fail to implement a policy of shared fiscal responsibility, they also largely failed to force the now self-governing colonies even to defend themselves. In 1860, Sir Charles Adderly rose in the House of Commons to complain that while Great Britain contributed £4 million to empire defense, the colonies contributed less than £40,000. "Why," he asked, "should the colonies be exempted from paying for their own defense? . . . it was," he continued, "absolutely unparalleled in the history of the world that any portion of an empire— colonial provincial or otherwise—should be exempted in purse or person from the costs of its own defense as was a British colony." Even that arch imperialist, Benjamin Disraeli, exclaimed two years later: "These wretched colonies . . . are a millstone around our neck."

The imperial record was bleak indeed. When the Cape ceased to make its £10,000 annual contribution toward the upkeep of the British garrison stationed in the colony, the treasury could only reserve the right to reopen the subject at another time and lament, "My Lords cannot, however, leave unnoticed the implication . . . that a colony should not be called upon to contribute towards the cost of a force maintained as an imperial garrison within its territory, with this opinion Milords cannot agree."

The £10,000, however, appears almost minuscule in comparison with the costs of that great "imperial conflagration," the Boer War, but so do the contributions of the self-governing colonies. It is estimated that the war cost the exchequer £216,166,000 and almost all of that sum was, in the final analysis, paid by the British taxpayer. Long negotiations for a substantial colonial contribution came to naught, and six years after hostilities ended, the British government conceded that it did "not propose to press for further payment. . . . " As Lord Sherbrooke, the chancellor of the exchequer, wrote: "Instead of taxing them as our forefathers claimed to do, we, in the matter of military expenditure, permit them in great degree to tax us."

Indeed, the colonists in the self-governing empire prospered while the British paid. In 1902, a British resident was contributing fifteen shillings (180 pence) annually toward imperial naval defense, Canadians paid nothing, settlers in Cape Colony paid three-and-a-quarter pence, in Natal, they paid five-and-three-quarters pence, in New South Wales, it was eight-and-a-half pence, and in Victoria and Queensland, it was one shilling each. In 1911, the Home government urged a small Canadian contribution to imperial defense, but with the certainty of moral rectitude, the government in Ottawa argued that the Dominion was already bearing more than its fair share of the burden. As proof, it pointed to its "Fisheries Protection Service, which cost £50,000 a year and maintained an armed boat on the Great Lakes which would soon be joined by a second on the Pacific Coast." "And," the Canadians righteously emphasized, "the entire enterprise did not cost the British taxpayer a farthing!"

The British should not have been surprised. This was, after all, the same Canadian government that, half a century before, with an American army poised on its border, had responded to a British request for a military contribution with the stirring response: "The best defense for Canada is no defense at all."

PUFFED-UP ZULUS, MEAN MAORIS, AND POOR JOHN BULL

Imperial defense was one thing, but the British fared only marginally better in their attempts to force the self-governing colonies to pay for their own protection. When the Zulu War of 1879 broke out, the costs were "temporarily" defrayed by the British Government. Peace came, but not repayment. Only one-fourth of the £1 million cost of the hostilities was ever squeezed from a reluctant Natal. At the end of the largely unsuccessful negotiations, the colony's governor complained to his superiors in the Colonial Office:

> This little puffed up council has command of the purse strings as fully as the House of Commons at home, and its members are more puffed up with an idea of their importance even than your English country member is.

Nor was the treasury ever able to gain more than token recompense from the Cape Colony for its share of the Ninth Frontier or Border War. After years of negotiation, the treasury admitted defeat and wrote, "My Lords will consent to accept £150,000 in full discharge of the debt." They may have been willing to accept, but the colony was certainly not willing to pay.

While one might argue that there was a small imperial component in some of the South African adventures, no such rationale can be used in the case of New Zealand's Maori Wars or Canada's problems with Luis Real's uprising on the Read River. After years of heavy fighting, hostilities in New Zealand ended as soon as the British "withdrew the legions." The colony's behavior prompted Lord Lyttleton, the colonial reformer and Gladstone's brother-in-law, to write, "the colonists have been so long carried in nurses' arms they cannot stand on their own feet. My belief is that this war against 2,000 aborigines has cost John Bull the best part of 3,000,000 [pounds]."

The Canadian case, too, presents a nearly perfect caricature of the state of political relations between Britain and her self-governing colonies. The home government was forced to pay for suppressing a rebellion aimed only at preventing the Dominion's westward expansion. Despite strenuous attempts, it was never able to gain any recompense; yet it continued to heavily subsidize the Canadian defense establishment.

REMONSTRATING VAINLY

In general, the stronger the representative institutions enjoyed by a colony, the better it was able to resist the demands and supplications of London. Additional evidence on the cost of empire, however, may be gleaned even from the record of negotiations between the British government and those colonies over which it could exercise absolute political control. In the case of so-called imperial levies, even colonies as insignificant as Mauritius and Bermuda were able to fight off the demands of the treasury. The government could do no more than tiredly note that it had "remonstrated vainly."

Regardless of the views of the treasury or the state of a colony's constitutional development, all colonies were advantageously positioned when it came to paying for actual hostilities. Wars were not infrequent along the imperial borders, and the initial expenses were almost always advanced to the local governor from the Treasury Chest—a fund of several hundreds of thousands of pounds spread throughout the Empire for use in emergencies. Once the imperial monies had been expended, the British usually enjoyed but little success in recouping their funds. For instance, despite dependent colonial status and an annual budgetary surplus that averaged about £20,000, Sierra Leone was able to avoid paying most of the costs of military operations conducted in the colony. Even the smallest victory seemed hard to gain. The fine of £2,400 that had been imposed on the offending monarch of Gambia would have effectively paid for the war, but the colony's governor had already remitted three-fourths of the mulct when the king signed a treaty guaranteeing free trade within his territory. "Milords are of the opinion," wrote the British treasury minister to the Colonial Office in 1862, "that no sufficient case has been made for our relieving the Government of Gambia from colonial expenses incurred in connection with the war against King Badiboo."

In 1860 Adderly had suggested that the financial assessment imposed on the British taxpayer by the colonies should no longer be countenanced; however, half a century later little had changed. The British enjoyed some success in gaining support from India, although that unique imperial entity received the protection of the British defense umbrella at bargain basement prices. Similarly, the government in London was able to extract some contributions from a few of the dependent colonies—Ceylon, Mauritius, and Hong Kong all made some contribution—and in one of its few unblemished triumphs, the treasury successfully cajoled the Straights Settlements into accepting fiscal responsibility for all of the costs of the Perak War of 1875. At some point, however, it must have become clear that, if the British wanted the vicarious pleasure of looking at a world map much of whose surface was painted red, the taxpayers in the metropole would have to pay virtually all of the costs. Discussing the impending assumption of control over the Transvaal, Sir Garnet Wolseley, Gilbert's "very model of a modern major general," had noted that "to enable us to hold our own we must be prepared to maintain a large garrison of British troops here, the expense of which must be defrayed by the imperial exchequer." To which Sir Michael Hicks-Beach, the colonial secretary, had roared in reply: "Now I must confess this is a position that can hardly be maintained." But it was.

THE CASE OF NATO

When the United States signed the North Atlantic Treaty in 1949, Americans believed they had initiated a policy of shared responsibility for European defense. From the very start, however, the burden fell heavily on the American taxpayer. As early as April 1950, *The New York Times* reported that "representatives of the smaller countries are frank in declaring that the plan's fulfillment depends on the continuation and perhaps increase of the United States' contribution to money and arms." The attitude of the larger countries was little different. The United States was determined that Western Europe should not fall prey to the Russians, and hence by 1951 accepted the inevitable, as had been the case with Britain and its Empire—the cost of European defense would be largely an American responsibility. The European per capita share of the European defense burden was only about one-third of the amount that the average American was asked to bear.

As the European Miracle of postwar economic recovery became increasingly dramatic, American policymakers came to expect that Europe would be willing to assume more of the responsibility for its own defense. Besides, the United States was faced with massive government deficits and serious foreign exchange problems. What seemed so obviously fair, however, was not to develop. In 1990, European per capita defense expenditures remained about where they had been in 1955, and the Europeans were no more interested in shouldering a larger fraction of their own defense costs than they had been at the time the North Atlantic treaty was signed. In November 1960, when Secretary of the Treasury Robert Anderson traveled to Germany to solicit a greater West German contribution to NATO, Bonn pleaded poverty. In words reminiscent of the Pennsylvania legislature's eighteenth-century response to Britain's call for help against the French, *The New York Times* of November 22, 1960, reported the West German government as asserting:

> [T]he total burden and hence the tax burden is at least as high in relation to the national income in West Germany as in the United States. Although welfare expenditures of various kinds make up a greater proportion of the German budget than they do in the American budget . . . such expenditures are necessary. Because of the country's position next to the Iron Curtain . . . welfare payments demonstrate that a capitalist system can prevent poverty.

Taken as a whole, the European nations, much like the British colonies in a previous era, were aware of the power of their position and politely resisted American attempts to raise the percentage of their gross national product devoted to defense. The colonies knew that the "mother country" would pay the costs of imperial defense if they did not; the "NATO allies" became convinced that the United States would never allow a militarily superior Soviet Union to be in the position to dominate Europe. Some hint of the strength of the American commitment can be gleaned from the following words of a famous politician:

> We have inaugurated an immense military program. The consequences of this program to our economic life are already evident. Under it, taxes will take a greater portion of the national income than that taken by the most noncommunist countries in Europe. Already we are in the midst of a disastrous wave . . . from its pressures. We must defer needed improvements. We can stand this for possibly two or three years pending a genuine rally in the noncommunist world to their

full part in defense. But we must in time have relief from a large part of the burden.

These words did not come from the lips of Jimmy Carter, Ronald Reagan, or George Bush, nor from Caspar Weinberger or Dick Cheney, but from Herbert Hoover speaking about the North Atlantic Treaty in 1950.

CONCLUSION

British statesmen recognized that empires cost a lot of money, though the burden was grudgingly accepted throughout most of the Victorian era. Americans in the twentieth century, burdened by the defense of Europe and Japan, have only gradually learned the same lesson. The nation that aspires to dominance must know, or will soon find out, that expansionism should be viewed as a great expense, not as a source of profit. Those who live under the cloak of protection spread by the dominant power, and benefit from it, will employ every means at their disposal to avoid sharing in the cost of the enterprise. Britain and its Empire, and the United States and its NATO allies, are but two examples of this phenomenon. Happy is the nation whose responsibilities are wholly domestic.

4

How Military Mobilization
Hurts the Economy

ROBERT HIGGS

Most people know that military mobilization is not cheap, but few appreciate the full range of economic consequences of going to war. President Eisenhower popularized the myth that war is good for business when he warned, in his farewell speech, of the dangers posed by a "military–industrial complex." It is still widely believed that Franklin Roosevelt lifted the country out of the Great Depression by getting us into World War II. The fact is, hasteful and wasteful military buildups have led to budget deficits, higher taxes, inflation, expensive transfer programs for veterans, and government intervention in markets—all of which have continued for decades after the wars they were supposed to facilitate.

Moreover, by diverting workers and resources to a bloated, privileged, anticompetitive procurement complex, war buildups have actually *reduced* the American capacity to invent, innovate, and enhance productivity along nonmilitary lines. America's competitiveness has thus diminished during recent decades as the Germans, Japanese, and others have concentrated their scientific and technological resources in commercial rather than military concerns. Military mobilization both causes an immediate resource drain and bequeaths to later generations legacies that alter the institutional structure, the scope of private property rights, and the economy's capacity to generate a rising output of consumer goods.

FISCAL FIASCOES

Consider the following figures. Total direct costs for major American military conflicts, expressed as a percentage of one year's contemporary gross national product, have been estimated as 104 percent for the Revolutionary War; 74 and 123 percent for the North and the South, respectively, in the

Civil War; 43 percent for World War I; 188 percent for World War II; and about 15 percent for both the Korean War and the Vietnam War. Costs associated with the Cold War have varied between 4 and 13 percent of gross national product annually during the "peacetime" years of the past four decades; they are currently accruing at less than 5 percent.

To finance these buildups, Uncle Sam has raised existing tax rates and, when this has not been enough, created new taxes. During the Civil War, for instance, the Union government jacked up tariffs, levied excises, and, for the first time in the country's history, taxed *incomes*. The war-spawned income tax lingered until 1872, and high tariffs remained for decades. The Sixteenth Amendment to the Constitution, ratified in 1913, cleared away constitutional barriers just in time for Congress to enact an income tax law on the eve of World War I. After the United States entered the war, Congress raised the rates enormously: Personal exemptions were reduced from $3,000 to $1,000 for a single person; the first-bracket rate was increased from 1 percent in 1913–1915 to 6 percent in 1918, and the top-bracket rate was raised from 7 to 77 percent. World War II witnessed similar tax changes. Personal exemptions were chopped and tax rates were raised severalfold: The first-bracket rate went from 4.4 percent in 1940 to 23 percent in 1944–1945, while the top-bracket rate rose from 81.1 percent on income over $5 million to 94 percent on income over $200,000. Payroll withholding of personal income taxes was instituted. The population liable for income tax payments increased more than threefold. Fewer than 15 million individuals filed income tax returns in 1940; nearly 50 million filed in 1945. Later conflicts have not given rise to tax increases as steep as those of World War II, but every war has produced tax increases. The lesson is that war finance and soak-the-rich schemes tend to coincide.

Nor has the government been eager to surrender its augmented revenues once war emergencies have clearly passed. Indeed, despite the close association of war and tax increases, U.S. governments have been hesitant to levy enough taxes to pay for the wars contemporaneously, and have shifted much of the burden away from current taxpayers by means of debt finance and inflation, leaving future taxpayers to finance war debts. The Revolution and the Civil War were notorious in this respect: Only about 10–15 percent of their direct financial costs were covered by current taxation. Wilson's administration got about 24 percent of its war expenses from tax revenues; Franklin Roosevelt's were somewhat more, running at about 41 percent. Only the Korean War, a relatively cheap mobilization, was fought on a pay-as-you-go basis—assisted, one must note, by price controls. When the Johnson administration delayed requests for a tax increase to finance outlays for the Vietnam War, it resorted to borrowing, which put pressure on the Federal Reserve System to ease monetary conditions so that the government's surging demand would not drive up interest rates abruptly. The monetary authorities accommodated the government's deficit by purchasing bonds in the open market, an action that increased commercial bank reserves, prompting them to make more new loans and investments and thereby increase the nation's money stock more rapidly. The consequence was an acceleration of inflation. Ultimately the Nixon administration, inheriting a difficult political choice, opted for mandatory wage–price–rent controls during 1971–1974.

As if higher taxes and inflation were not bad enough for the economy, the country has been haunted by other consequences of America's wars. Expensive transfer programs for veterans have been adopted, and the payments have continued for decades after the wars. In addition, the enormous increase in the government's budget deficit, associated with the Reagan administration's military buildup during its first term, continues to loom as an impediment to economic health.

BUREAUCRATIC BUILDUP

More destructive than the fiscal fiascoes have been the institutional developments. Even with debt finance, inflation, and taxation, the U.S. government has been hard-pressed to command the resources deemed necessary for successful prosecution of its wars; therefore, it has invariably turned to measures of command and control. Price–wage–rent controls, obligatory orders, physical allocations of materials, rationing, and priority assignments have been extensively used. These interventionist measures do not avoid economic costs of warfare; they simply obscure them, and thereby blunt the opposition of those who bear the costs.

Worse yet, this domestic interventionism has fostered increasing governmental control of the economy. Though private contractors play an integral role, mobilization has never been a genuine market phenomenon: it has been a politicobureaucratic planning process. Military buildups encourage governmental takeovers throughout the economy as special interests, whose activities are only spuriously related to the military strength of the nation, press their claims under the rubric of "national emergency" or, during the Cold War, "national security."

As has been the case with war financing, interventionist inroads have proven resistant to change even after the various crises have passed. The peacetime draft introduced in 1940 remained, with only a brief hiatus in the late 1940s, until 1973. Furthermore, since 1940 the military procurement system has constituted a voracious central-planning apparatus embedded in the (potentially) most dynamic "high-tech" sector of the American economy. The military–supply business is today, as in past crises, the most heavily regulated in the economy. Excruciatingly detailed rules for the conduct of weapons contracting fill more than 1,850 pages in the three volumes of *Federal Acquisition Regulations*. The defense business, at best a twisted and mutant form of free enterprise, is especially vulnerable to political manipulations—for example, schemes in which members of Congress buy votes by generating local defense-related employment. These political interventions sap the efficiency of the defense sector and distort the overall economy in which it operates.

HASTE MAKES WASTE

Additional costs have been borne because the United States has entered all of its major wars unprepared. Because of inadequate planning, war programs have been implemented in extreme haste—and it is a basic economic theorem that, other things being equal, completing a fixed-volume, fixed-rate production program at an earlier date costs more than completing it at a later date. Jesse Jones, a leader of the government's efforts to build up the industrial base for war production during World War II, recognized this relation. "We had to hurry," he observed. "Haste, of course, added to the cost."

Many people now look back to World War II as an occasion when the military–industrial complex operated with clearer purpose and greater effectiveness than it does presently. This view is probably correct. Historians, however, have concluded that even during World War II, the production program was a "buckshot operation." It was a classic case of trying to solve a problem by "throwing money" at it. In today's dollars, the cumulative war expenditures of 1940–1946 amount to more than $2.8 trillion. The average annual spending over this seven-year period was about $400 billion, or near 50 percent more than the current annual defense spending—and this amount was extracted from an economy less than one-third as large as ours is today.

Later buildups have been less demanding of resources, but each has exhibited the same pattern of surprise, hasty mobilization and enlarged cost. Even the Reagan buildup, large in absolute cost if small in its increased share of GNP, was a hurried, almost frenzied affair. In the exhilarating early days of the Reagan administration, budget director David Stockman thought the costs of the buildup could be moderated by draining what he characterized as "a kind of swamp of $10 to $20 to $30 billion worth of waste that can be ferreted out if you really push hard." He later conceded, however, that "the defense numbers got out of control and we were doing that whole budget-cutting exercise so frenetically, so fast, we didn't know where we were ending up for sure." In essence, the Department of Defense (DoD) got a "blank check."

Stockman thought at the time that the DoD had "got so god-damned greedy that they got themselves strung way out there on a limb"—one that inevitably would be chopped off. He was wrong. In fact, the DoD steadily enhanced its position during Reagan's first term, which saw an increase of over 40 percent in real defense budget authority, and it almost held its own in the reaction that occurred during the second term. When the post–1985 slowdown came, however, the Pentagon had committed itself to a host of new weapons programs. Rather than cancel a significant number of them, the defense managers chose to pare the production runs, keeping almost all programs alive. Such stretchouts resulted in accelerating unit costs and diminished efficiency of overall defense procurement—literally, less bang for the buck. Between 1977–1981 and 1982–1985, for example, Grumman's annual production of F–14 fighter planes fell almost 30 percent, and the real unit cost of the aircraft rose by 21 percent. In short, it would be incorrect to associate the increased production during a military buildup with an inherent increase in productivity or efficiency, as is often done.

THE PROCUREMENT OF PRIVILEGE

The procurement system that now forms the heart of the military–industrial complex has obvious roots in the mobilization of World War I, when hundreds of businessmen transferred from their firms to government boards to tell the Wilson administration how to organize its war production program. In its modern form, however, the system springs most clearly and continuously from events in the second half of 1940. The form it took reflected the usual pattern of inadequate planning and preparation for war, surprise, and hasty improvisation.

After the astonishing German advances in the spring of 1940, the U.S. government decided quite suddenly that national security required an enormous buildup of the military establishment. After twenty years of starving the army and the navy, Congress opened the budgetary floodgates. Between June 1940 and December 1941 about $36 billion was made available to the War Department alone—more than the army and navy combined had spent during World War I. As Secretary of War Henry L. Stimson remarked, however, "the pinch came in getting money turned into weapons." The nation possessed enormous potential to produce munitions, but its munitions industry was diminutive early in 1940. According to Donald Nelson, later the head of the War Production Board, it was "only a token industry." The rearmament program somehow had to "enable American industry to make the heavy capital commitments, plant expansion, and organizational changes essential to large-scale armament production."

The government's response to its predicament proceeded on the basis of Stimson's Axiom: "If you are going to try to go to war, or to prepare for war, in a capitalistic country, you have got to let business make money out of the process or business won't work." To assure businessmen that they would profit sufficiently—and without much risk—the administration sponsored and the Congress passed several important statutes in summer and fall 1940. These acts enlarged the ability of the military departments to enter into negotiated, as opposed to competitively bid, contracts; to enter into cost-reimbursement, as opposed to fixed-price, contracts; and to make substantial advance and progress payments to contractors.

A dramatic shift of contracting practices ensued. In fiscal year 1940, the War Department had made 87 percent of its purchases through advertising and invitations to bid, but during the eight months after June 30, 1940, when the department spent ten times more, it placed 74 percent of its contracts by negotiation. Although procurement officers were not allowed to use cost-plus-a-percentage-of-cost contracts (CPPC), which had caused scandals during World War I, they could and did use cost-plus-fixed-fee contracts in which the fixed fee was normally figured as a percentage of estimated cost. (Profits for cost-plus-fixed-fee contracts are still established in the same way). Thus, as John Perry Miller concluded from an exhaustive study, "we came closer than is generally realized to financing this war on a disguised CPPC basis."

The government took much of the sting out of high wartime tax rates by allowing the contractors rapid (five years or less) amortization of investments in war facilities. They could also "carry back" certain postwar costs and losses by claiming refunds against excess-profit taxes paid during the war. To shift even more risk off the contractors, the government itself financed two thirds of the investment in manufacturing plant and equipment during the war. Most of the governmentally financed plants were operated on liberal terms by private contractors who were given privileged standing to acquire the facilities after the war. Hence risks of capital loss again were sloughed onto the taxpayers.

Although comprehensive data are not available, existing evidence shows that the war-supply business was extraordinarily lucrative. A study of 3,178 corporations whose profits were renegotiated during fiscal year 1943 found that, for contracts with the government, corporate rates of return on net worth, after taxes and renegotiated refunds, ranged from about 22 percent for the largest firms to 49 percent for smaller firms—extraordinary profits given that the contractors bore little or no risk.

Large manufacturing firms enjoyed the bulk of the business. The top 100 prime contractors received about two-thirds of the awards by value; the top 10 got about 30 percent; the leading contractor, General Motors, by itself accounted for nearly 8 percent of all prime contracts by value. Research and development (R&D) contracts with private corporations were even more concentrated. The top sixty-eight corporations got two-thirds of the R&D awards; the top ten firms took in nearly two-fifths of the total. Here was a harbinger of how defense R&D would be allocated during the postwar arms race driven by scientific and technological competition with the Soviet Union. The concentration of governmentally financed facilities was even greater than the concentration of war production, prime contracts, or R&D awards. Just twenty-six firms enjoyed the use of exactly half the value of all governmentally financed industrial facilities leased to private contractors as of June 30, 1944. The top 168 contractors using such plants enjoyed the use of more than 83 percent of such facilities by value. The implication of this high concentration for the character of the postwar industrial structure is evident when one recalls that the operator of a government-owned contractor-operated facility usually held an option to buy it after the war.

The arrangements created in 1940 and refined during the next five years produced a complete transformation of the relations between the government and its military contractors. In the words of Elberton Smith, the official army historian of the mobilization, the relationship

> was gradually transformed from an "arms length" relationship between two more or less equal parties in a business transaction into an undefined but intimate relationship. . . . Contracts ceased to be completely binding: fixed prices in contracts often became only tentative and provisional prices; excessive profits received by contractors were recoverable by the government; and potential losses resulting from many causes—including errors, poor judgment, and performance failures on the part of contractors—were averted by modification and amendment of contracts.

Smith's description of the military–industrial business during the war applies equally to the present-day business. Now as then, the procurement business—at least where the major contractors are concerned—is open, fluid, and subject to mutually beneficial adjustment. A transaction is less a firm "deal" than an ongoing joint enterprise among colleagues and friends—as Smith put it, an "intimate relationship"—in which military officials and businessmen cooperate to achieve a common goal not incompatible with, but rather highly facilitative of, the pursuit of their separate interests.

Richard Stubbing, a veteran defense analyst at the Office of Management and Budget, has observed that "contractors have not been held seriously accountable for the cost figures they cite in their competitive bids." The Pentagon, not wanting its contractors to go bankrupt, occupies "the conflicting position of wanting to hold defense firms to their contracts while at the same time wanting to protect them." During the past twenty years, Pentagon bailouts of Lockheed, Litton, General Dynamics, Chrysler, Grumman, and other leading contractors have demonstrated that the propensity to protect contractors outweighs the inclination to hold them to the terms of their contracts.

Postwar defense profits have reflected the contractors' privileged position. A DoD study of major prime contractors during the period 1970–1983, a time of famine followed by feast, ascertained that, on the average, military work generated returns on assets of 20.5 percent—about 54 percent higher than the 13.3 percent earned on investments in all U.S. durable goods manufacturing. In 1982 and 1983, the most recent years considered in the study, the defense firms achieved rates of return exceeding 25 percent.

CONCLUSION

Despite the disintegration of the Soviet Union, Russia continues to maintain a military capability of awesome dimensions. In the event that

Russian–American relations become truly and enduringly cordial, a variety of other threats to U.S. interests will persist. Whatever the strategic doctrine embraced by the U.S. government, continuation of a substantial national defense program seems assured. Most Americans would agree that an adequately protective program conducted at minimum sacrifice of our resources is desirable. As we carry forward in search of such a program, what does a study of the past teach us that might help us to avoid some of the old mistakes? Perhaps the most obvious lesson is that it is a good idea to avoid "crash programs." Certainly the neglect of the military establishment that preceded both world wars contributed greatly to the difficulties experienced during those buildups.

Of course, since the Korean War the military establishment has not been allowed to sink anywhere near the depths reached during the interwar period. Still, the plunge of 1968–1978, when the national defense share of GNP dropped from 9.0 to 4.8 percent, followed by the buildup to a share of 6.6 percent in 1986, shows that considerable instability still plagues the effort. Both the efficiency of the defense sector and the effectiveness of military procurement planning would benefit from a steadier and more predictable flow of defense spending. In general, however, the ills that afflict the operation of the military–industrial–congressional complex are too deep-seated and complicated to yield to any simple treatment; indeed, they are probably incurable as long as the defense effort continues on anything like its present scale.

Marginal improvements, however, might be made—and they might save tens of billions of dollars. History suggests that avoiding cost-reimbursement contracts would probably help to contain costs. Contracts could be written so that contractors receive a reward for bringing their products to the military on schedule and at or below the costs estimated at the outset; contractors could be penalized for failure to adhere to commitments. Even more helpful than writing the optimal *form* of contract would be a restoration of the binding force of the contracts, whatever their form. When defense contracts are no more than the legal outer garments of a fluid and easily adjusted "intimate relationship," it hardly matters what the form of the contract may be. Past experience also suggests that putting more effort into independent monitoring of contractual compliance might yield a high rate of return to the taxpayers. There is, however, no magic to exorcise the diverse demons that dwell within the military–industrial–congressional complex. As long as the United States continues to devote such enormous resources to the numerous and intricate enterprises of national defense, citizens will simply have to put up with bloated costs and substantial waste, fraud, and mismanagement. History teaches that lesson, too.

II

WORKERS AND EMPLOYMENT

Adam Smith, the Scotsman who invented economics two centuries ago, announced in the first line of *The Wealth of Nations* that "the annual labour of every nation is the fund which originally supplies it with all the necessities and conveniences of life." We are liable to be dazzled by financial wonders and forget his simple point: in the sweat of our brows shall we eat bread.

The story of American economic life must at bottom be a story of laboring, in office and marketplace, in field, factory, and mine. Thus, the next offerings: Fishback on death in the mine, Alston on bankruptcy on the farm, and Hoffman on technological unemployment at the factory. No wonder economics is called the Dismal Science.

The economist says, look below the surface. A good way to eliminate technological unemployment is to outlaw it, right? Wrong, unless you want a stagnant society, poor into the next century. A good way to help the farmers is to subsidize output, right? Wrong again. A good way to make the health of miners better is to give them workers' compensation for injuries, right? No, no, no: still wrong after all these minutes.

Adam Smith said it first and best, in Book I, Chapter 10: "The whole of the advantages and disadvantages of the different employments of labour and stock must, in the same neighborhood, be either perfectly equal or continually tending to equality." Labor and capital, like water, find their own levels, or they would "at least . . . in a society where things were left to follow their natural course, where there was perfect liberty." The attempt to stop the tide abridges the liberty without in fact stopping the tide, in field, in factory, and in mine.

5

Does Workers' Compensation Make for a Safer Workplace?

PRICE V. FISHBACK

We would like to live in a world free of accidents. Unfortunately, we do not. Our concern with accidents and their consequences seems to have increased over the past few years as newspapers have frequently reported on lawsuits for damages from plane crashes, product malfunctions, medical malpractice, workplace accidents, even tearing down football goalposts or being hit by a foul ball at a baseball game. In recent decades there has been a legal revolution in liability law that has increased the share of the victim's accident costs paid by employers and producers. While changes in liability law might aid the unfortunate victims of accidents, they may also lead to unintended consequences, like a sharp rise in product prices or a curtailment of some activities. Some claim, for instance, that a rise in the cost of malpractice insurance has led to a shortage in obstetricians to deliver babies. We can learn more about the unintended consequences of liability reforms by studying an earlier revolution in liability: the introduction of workers' compensation for workplace accidents.

COMMON LAW AND COMMON SENSE

In the late 1800s, before workers' compensation, liability for workplace accidents was based on common-law standards of negligence. If a worker was injured on the job, the employer was not expected to pay the injured workers' costs from the accident unless the employer had failed to exercise "due care." Due care under the law meant that the employer followed several practices. First, the employer offered and enforced reasonable safety rules, and posted warnings of dangers. Second, enough qualified workers to handle the job were hired. Third, the tools customarily used by prudent people on the job were provided. In a coal mine, for example, the employer provided adequate numbers of props to keep the roof supported. The tools provided by an employer did not have to incorporate the latest safety advances unless they were in widespread use. Thus, when a new chainguard for a coal cutting machine was invented, the employer who did not provide it could still be exercising due care, as long as the chainguard was used in only a few mines in the industry. Nor was an employer liable for accidents caused by the worker using a tool in uncustomary ways, such as riding a mine car designed only to haul coal—particularly if the employer had posted rules against it. The logic of negligence liability was to force the employer to pay for accidents that might reasonably be expected to be prevented.

Injured workers faced some additional legal obstacles to compensation for their injuries. Even when employers failed to meet their safety responsibilities, they could employ any of three legal defenses: assumption of risk, contributory negligence, and the fellow-servant doctrine.

Under assumption of risk, employers could ask workers to show that their accidents were not caused by factors that were ordinary for that type of work, or, if extraordinary, that the risks were not known and accepted. A steeplejack who tripped and fell off of a steeple might not have received compensation from an employer because such risks were known and accepted when the steeplejack took the job. Typically, workers demanded a higher wage as compensation for these dangers before accepting the job.

Under contributory negligence, workers could not collect damages if they might have avoided the accident by exercising due care. The employer was probably not going to be liable for injuries to a motorman who slammed into a wall while driving too fast to make a turn. This doctrine led to lower-cost accident prevention by giving the motorman incentives to drive safely. Of course, the employer might have prevented the injury by padding the walls at every turn—an expense possibly less than the expected costs of the accident—but the motorman could prevent the injury at even lower cost by driving at a proper speed.

Finally, under the fellow-servant doctrine, an injured worker was not compensated if the accident had been caused by the actions of another worker. A miner was not likely to be compensated by an employer if a partner's failure to correctly prop the roof caused injury in a roof fall. This defense helped prevent accidents by giving workers added incentive to report careless fellow workers who might cause them harm.

At first look, the employers' defenses in accident cases seem impregnable, which has led some to argue that the laws were designed to favor entrepreneurs over workers in trying to stimulate economic growth. What incentive did employers have to prevent accidents when it seems they rarely had to compensate anybody? In fact, roughly 30–70 percent of the families of fatal and serious accident victims received some form of compensation, usually about one year's earnings. Most compensation came in the form of out-of-court settlements within six months of the accident, avoiding the long delays, sometimes four to six years, associated with seeking a decision in the courts.*

More careful consideration shows that negligence liability and the three defenses encouraged common-sense prevention of accidents by the parties with the lowest costs of prevention. The courts assigned employers liability for accidents where they were the lowest-cost preventers. In coal mines, for example, mine employers expected to have to pay settlements when accidents resulted from failures to provide adequate mine ventilation, proper precautions in haulageways, mine gas inspections, watering of coal dust to prevent the spread of mine fires and explosions, and adequate numbers of timbers for miners to prop the roof. Workers, for their part, had incentives to prevent the remaining accidents, which took place primarily in rooms where they were the lowest-cost preventers, because they received no compensation if injured.

THE HAZARDS OF SAFETY

The major shift in liability for workplace accidents came with the passage of workers' compensation laws in most states between 1910 and 1930. The statutes replaced negligence liability with a form of strict liability, under which the employer was obligated to pay employees or their heirs a fixed amount for *all* serious accidents "arising out of employment." Further, the amounts of compensation rose. Compensation for fatal accidents rose from roughly a year's income in settlements under negligence liability to a minimum of three years' worth of earnings under workers' compensation.

Advocates of workers' compensation predicted that an increase in payments to injured workers would force employers to prevent more accidents, and that accident rates would therefore fall. As employers paid more to injured workers to offset the costs of their accidents, however, workers had less incentive to prevent those accidents from happening. In the coal industry, fatal accident rates increased after workers' compensation laws were passed. For every 1,000 men who worked full time in bituminous coal mines in the United States from 1900 to 1930, between 3 and 4 men were killed in accidents. The results of one recent study suggest that passage of workers' compensation laws raised the fatal accident rate in the coal industry nearly 20 percent.

The rise in payments to injured miners with the passage of workers'

---

*Both sides tended to use the threat of court delays to their advantage: Some injured minors were able to get larger net settlements than they would have otherwise, while some employers were able to get away with paying lower settlements to injured workers. After the switch to workers' compensation, the new methods of administering compensation were much faster than use of the court system had been, but it is not clear that they were much faster than settling negligence suits out of court. The switch to workers' compensation may have reduced the payments going to lawyers, but even that is not clear, since legal representation was still useful to workers and employers in workers' compensation hearings.

compensation gave coal employers more incentives to prevent accidents, but the increased compensation reduced the incentive for miners to work carefully. Since coal loaders and pick miners were paid by the ton of coal, they saw that by working a little faster and taking more risks they could get higher earnings—even though a roof fall injured or sometimes killed miners who tried to finish loading cars before setting new props for the roof. Under negligence liability, miners had extra incentive to work more slowly and safely because if they got injured in a roof fall, they were likely to get little or no compensation. Under workers' compensation, however, they could take more risks because they were assured of compensation that was often higher than what they would have received under negligence liability.

This increase in risktaking led to more accidents like roof falls. The problem was that roof falls in the miner's workplace were the types of accidents that employers were not very effective at preventing. Effective prevention might have required hiring a large number of supervisors to watch the miners. The costs of hiring supervisors to prevent the extra accidents caused by moral hazard appeared to be higher than the expected damage payments. Rather than take these costly steps, the employers chose instead to pay the extra damages.

The move to workers' compensation did not raise accident rates in every industry. Fatality rates for machinery accidents in industry actually fell, and it appears that the efforts of manufacturers to reduce accident rates more than offset moral hazard problems. The difference between the results in coal mining and in manufacturing probably can be traced to a difference in the cost to the employers of preventing accidents. Supervisors could monitor the workers' use of machinery more easily in manufacturing than in coal mining, and companies discovered that they could eliminate many machinery accidents by putting more footguards and handguards near whirring blades and gears. In contrast, many of the accidents in the miner's room in a coal mine were a result of natural conditions that the employer could not fix at low cost.

THE WAGES OF DANGER

The introduction of workers' compensation also affected the wages paid to coal miners. Wages in labor markets often rise to offset workplace dangers, but will readjust downward when the dangers are reduced. Coal miners received higher hourly earnings than other workers in part because mining was more dangerous than other work. This wage premium was relatively high under negligence liability because so many accident victims received small amounts of compensation. When workers' compensation raised the postaccident compensation for accident victims, the coal companies were able to lower the wage premium paid for accident risk and still attract an adequate number of workers. In essence, workers' compensation gave more money to the victims of accidents, but it took away from the wages paid to all workers.

CONCLUSION

The tale of workers' compensation offers some lessons for current discussions of liability reform. Liability laws determine who pays for an accident—the victim, or another person or institution. Thus, the laws establish the incentives for people to prevent accidents. A change in liability can shift the burden of accident-prevention. As we saw in the coal industry, this can lead to the unintended result of higher accident rates. Furthermore, the liability laws influence a wide range of other economic variables, like prices and wages. Paying more in compensation to accident victims may come at the expense of other workers. In the early coal industry, the gains that injured miners received from higher compensation were offset by a reduction in hourly earnings for all miners. The trend in liability law over the past few decades followed a similar path of having other parties pay an increasing share of the injured person's accident costs. Workers' compensation has become increasingly generous with postaccident compensation, and some studies with recent data have shown that wages have consequently fallen and accident rates risen. As court decisions have stripped away the defenses available to producers in product liability cases, the trends have imposed costs on the average consumer in the form of higher prices and curtailed production of many products like ladders, football helmets, prescription drugs, and medical services. While we seek the laudable end of helping unfortunate accident victims, when making public policy, we must also recognize the wide-ranging and sometimes unintended costs imposed on other members of society.

6

American Farming:
If It's Broke, Why Can't We Fix It?

LEE J. ALSTON

With all the media attention to the plight of American farmers in the 1980s, one might have thought that farms had never failed before. In fact, things have been much worse. In the 1920s and 1930s, agricultural distress prompted the federal government to take action to prevent farm failures. The current policies under review—government payments to farmers under existing commodity programs, infusion of capital into the Farm Credit System, and state moratoria on farm foreclosures—are extensions of programs originated in the 1930s. Perhaps, then, we can learn from past experience.

THE PRICE OF FAILURE

Throughout the 1920s and 1930s, one out of fifty farms failed each year—close to 100,000 annually, or 2 million over the two decades. It is not too surprising that farms failed in the 1930s: times were tough for everybody. Why then did farms fail in the roaring 1920s?

Essentially, they failed because times were too good in the 1910s. Farm prices skyrocketed during World War I and remained high after the war. By 1920, prices had more than doubled over the prewar levels. Fueled by optimism, farmers expanded their holdings and accumulated debt. In the more arid regions of the northern plains and the Southwest, farmers put to the plow land that had previously been used for grazing. Just as a football team plays its worst players last, the land brought into cultivation last was less productive than land already cultivated; however, with prices continuing to increase, nobody cared. In prime areas like Illinois and Iowa, all the land was already under the till, yet this did not prevent farmers in those states from participating in the boom. In their thirst for more land, farmers simply bid up the price. Easy credit greased the wheels of expansion. And why not? Crop and land prices just kept going up.

The agricultural party ended in July 1920. In the next year prices fell 30 percent, never to recover in the 1920s. As bleak as this seemed at the time, worse was on the horizon; prices fell another 50 percent by 1932 and never reached the level of the 1920s, much less the nirvana prices of the 1910s. Not only did prices fall, but boll-weevils periodically ravaged the cotton crop in the South, and droughts plagued farmers in the Great Plains and Mountain states.

When farm earnings fell, farmers sought help from the government. In 1928 Congress approved a two-price system that sought to increase earnings to farmers by paying them a higher-than-world-market price for domestically sold crops, livestock, and fiber. Unfortunately for farmers, President Calvin ("silent Cal") Coolidge vetoed the legislation. Herbert Hoover, who followed Coolidge to the presidency, had made agricultural policy a top priority in his campaign platform; once elected, he even followed up on his promises. With Hoover's support the Agricultural Marketing Act, creating the Federal Farm Board, became law in 1929. The Farm Board was endowed with $500 million, not a small purse in those days, and supported prices by offering loans with forthcoming crops as collateral. Whenever prices were high, farmers would sell their crops on the market and repay the government loan. If market prices were below the loan price of the government, farmers would give their crop to the government at the loan price.

Unfortunately for the government, market prices fell from the already low levels existing at the beginning of the price support program, and the Farm Board exhausted its funds by 1931. Beginning in 1929, evidence suggests that the Farm Board may have moderated the price slide, yet even this possible benefit would have been outweighed by the forced liquidation of accumulated stocks after 1931, which only depressed prices once more.

ROOSEVELT'S REMEDY

Franklin Delano Roosevelt was elected to the presidency in 1932 with a mandate to "do something" about the worsening plight of many Americans, including the farmers. His remedy for farm distress was twofold: (1) price supports coupled with restrictions on acreage, to raise farm earnings; and (2) renegotiation of credit terms, to relieve the burden from mortgage debt. The government supported prices by paying more than the world market price on the percentage of output consumed domestically. Participating farmers also received rental payments for land that was "set aside." Set-aside acreage reduced the supply of crops on the market, and this in turn propped up prices. These subsidies made a difference: In the early

New Deal years, rental and price-support payments made up as much as one-third of the income of some participating farmers.

Credit subsidization, the other half of the farm program, was exercised through the existing federal land banks. The land banks refinanced their own loans, as well as many held at private banks and insurance companies. It was a good deal for debtors: Refinancing meant a reduction in principal and lower interest rates. Initially, the federal land banks also refrained from foreclosing on delinquent farmers. Congress went even further, opening the public purse to debt-burdened farmers and authorizing the Land Bank Commissioner to make loans to farmers in severe financial distress. Since many farmers in this group were past saving, this turned out to be a real charity toss.

Naturally, farmers responded with open arms to the government largess. Although the government could not meet all comers, its share of farm mortgage debt increased threefold from 1933 to 1940. The statehouse got into the action, too. Between 1932 and 1934, twenty-five states passed moratoria on farm foreclosures, varying from three months to four years. Legislators hoped that earnings would rise sufficiently over the moratoria to enable farmers to make past-due mortgage payments. The moratoria were controversial, however, and were fought by creditors all the way to the Supreme Court. In a historic decision, the majority upheld a moratorium in Minnesota, arguing that "the economic interests of the state may justify the exercise of its continuing and dominant protective power notwithstanding interference with contracts." The court seemed to be saying that in times of emergency "anything goes." As Justice Sutherland dissented:

> He simply closes his eyes to the necessary implication of the decision who fails to see in it the potentiality of future gradual but ever-advancing encroachments upon the sanctity of private and public contracts. The effect of the Minnesota legislation, though serious enough in itself, is of trivial significance compared with the far more serious and dangerous inroads upon the limitations of the Constitution which are almost certain to ensure as a consequence naturally following any step beyond the boundaries fixed by that instrument.

States that did not pass moratoria frequently passed other laws easing the burden of debtors. Some legislatures lengthened the period over which a delinquent debtor could redeem property from a creditor. In Georgia, the redemption period was up to ten years. With such legislation, creditors had little incentive to foreclose.

EXTRA CREDIT

The farm sector was in dire straits throughout the depression, yet things did improve after 1933. Between 1933 and 1937, farm failures fell by half, from close to 4 percent to less than 2 percent. The decline was in part the result of federal and state agricultural policy. Recent estimates indicate that the combined government programs prevented approximately 200,000 farms from failing. This represented about 4 percent of all farms or roughly the entire failure rate for 1933. The moratoria on farm foreclosures, however, made existing loans more costly for creditors. Even though a farmer owed a banker money, the banker could not get it back by forcing the farmer to sell. To a certain extent, the banker's loss was the farmer's gain— and nobody felt sorry for bankers.

Still, the game did not end there. Bankers soon compensated for the inability to foreclose by charging higher interest rates on the mortgage or rationing credit. Evidence from the 1930s indicates that rationing of credit was the most frequent response to moratoria. Given the temporary nature of the moratoria, creditors may have reasoned that the short-run advantage of raising interest rates was outweighed by the adverse effects on their reputation. This was particularly true for eastern insurance companies, traditionally suspected of chiseling trusting Midwesterners. Once the moratoria and the Depression passed, creditors wanted to make sure that they had not lost all their borrowers.

The federal credit policies in the 1930s also imposed costs on taxpayers and private creditors. Thanks to subsidies from the U.S. Treasury, the federal land banks and the Land Bank Commissioner offered mortgage loans at lower interest rates than did private creditors. The subsidy was significant, amounting to at least a percentage point. Also having to compete with the federal government for loans no doubt raised the costs of doing business for banks and insurance companies. To some extent, private financial institutions were compensated by the government's willingness to take some bad loans off their hands; nevertheless, costs did increase—and some of those costs were borne by depositors.

The combined costs of the moratoria and subsidized credit appear small, however, when compared to the costs of increasing farm earnings. Between 1933 and 1935, direct government payments totaled over $1 billion and represented as much as one-third of farm income for some crops. This $1 billion included straight subsidies and rental payments for acreage set aside, but did not include the higher prices consumers had to pay for food and for cotton clothes. From 1933 to 1935, these indirect costs to consumers increased the farm bill to consumers/taxpayers by over $300 million.

WILL WE NEVER LEARN?

In the 1980s, farms failed at the highest rates since the end of the Depression. As in the 1910s, crop values skyrocketed in the inflationary 1970s, and farmers entered a bidding frenzy for land; for a while, even Arab investors got into the act. Farmers expected high prices to continue into the 1980s, as did the secretary of agriculture, who exhorted them to plant "fence row to fence row." When crop prices began to fall in 1980, troubles began. The U.S. government stopped collecting data on farm failures that very year, but recent estimates of average failures for the period 1981 to 1987 range from 3 to 4.5 per thousand farms. This may seem mild compared with failures in the 1930s, yet is significant when we remember that failures have otherwise exceeded 2 per thousand only *once* since 1943.

Today's farm policy, and the problem it seeks to solve, likewise has its parallels in the 1930s. Although state legislators have not yet reenacted the farm foreclosure moratoria of the 1930s, there have been fierce attempts in the mid–1980s to resurrect them. In Iowa, the state government imposed a moratorium by executive order. Farm credit is still subsidized; the Farm Credit System is known as the lender of last resort, with lending criteria that would give private bankers heart attacks. Last, but by no means least, are the subsidies designed to raise farm income. In 1986, these subsidies exceeded $25 billion, representing fully *one-third* of net farm income (!). Government payments did fall off somewhat in 1987 and 1988, but were up again in 1989.

All this preferential treatment is afforded our roughly 2 million farmers, who comprise less than 1 percent of our total population. As a consequence, the other 99 percent of us pay a lot more for our food because we pay farmers not to grow crops on all the available land. Estimates for the late 1970s indicate that consumers paid over $5 billion more per year for agricultural goods than they would have paid without government programs.

From this it is clear that, under our current farm policy, taxpayers and consumers lose, while farmers gain. Some would therefore say that, on balance, society as a whole is the same; however, this is wrong. There is a net loss to society, which diverts resources to farmers and, so far from receiving any benefit, actually receives higher food prices in return. Poor families, of course, are hurt worst of all; even those who do not actually pay market prices are victimized, since their food stamps go less far.

Given the past and present billions spent on subsidies, it is obvious that, sentimental–esthetic objections notwithstanding, we have too many farmers. If the price of crops were allowed to fall to market price, farmers— or, at least their offspring—would be compelled to move from, say, Mahomet, Illinois, to Chicago, and there to learn or practice a new career— much as steelworkers in Pittsburgh managed to do over the past decade. This does not mean that it would be costless for farmers to move from Mahomet to Chicago, or that there would be no trauma associated with changing occupations, however, it would, among other benefits to society, mean a reduction in the cost and trauma associated with being a poor parent, for example, feeding a family in the Chicago slums.

Our present surplus of farmers is in part the result of a spectacular increase in farm productivity. In fact, farm output has increased faster than demand, and it takes time for farmers to migrate to new jobs—but migration is slowed down as long as it pays to stay in farming. We all know who pays for farmers to stay in Mahomet: among others, it is the poor families in Chicago slums. The question is why we allow this kind of mis-allocation year after year.

The answer is straightforward: Farmers are a potent lobbying force. Unfortunately, lobbying is not just a matter of robbing consumers to pay farmers; consumers must also pay a service charge for the privilege of being robbed. Lobbying to maintain or capture subsidies adds to the cost of farming and farm produce, and to the waste of the government programs. It would be cheaper for us if no one paid to ask the government for its programs. It would be cheaper for us if no one paid to ask the government for *anything*, and simply taxed some folks and gave the money to other folks. Cheapest of all, of course, would be to let small family farmers— like all other types of small businesses in this country—go out of business when they cannot make it in the marketplace.

WHAT SHOULD BE DONE

It is obvious that we cannot afford ever-increasing support to agriculture. Many in the Bush administration advocate a return to free-market farm-ing—but then, so did many in the Reagan administration, during which subsidies reached all-time highs. As long as agricultural lobbies remain well-financed, and as long as public sympathy for agriculture remains high, we will continue to subsidize agriculture despite the costs.

Public sympathy for agriculture remains high because we care about the family farm. Ironically, without subsidies, the capital requirements for entry into agriculture would fall, presumably to the benefit of the small family farmer. Even if family farms were to give way to partnerships and corporate farms, this would only result in greater productivity. In the days before mechanized and science-based agriculture, small-farming units meant greater incentives for effort, as well as a reliance on local knowledge; farmers' fathers and grandfathers knew from past experience what sections of which lands were most fertile, and fertilized accordingly. With the aid of science, however, and even the nascent prospect of genetically engi-neered crops, the "oral tradition" is not as important as it was in grand-father's day.

Neither is it as difficult as it once was to monitor farm work. Before mechanization, it was impractical to motivate and monitor labor unless the land was cultivated by family members. For example, until the mechanical cotton picker spread through the South in the 1960s, the institution of sharecropping—which gave farm laborers a share of the output—remained as a partial solution to monitoring problems. Today, farming is much more standardized, and agricultural workers have much less leeway in adjusting their efforts. If a farmhand plows the back forty (today, the back 400 or 4,000) acres inside an air-conditioned John Deere tractor, listening to his favorite rock group on his CD player, he has much less discretion—or inclination—to shirk than he would if the plowing was being done behind a mule in 90° heat, with humidity to match. Farms are becoming like factories—standardized. To this extent we will not be sacrificing productivity—or worker comfort—if the family farm should disappear.

If it were not for government subsidies, family ownership of farms might not make economic sense. Farms are worth a lot of money, yet family farmers are seldom diversified in their assets. Selling off part of the farm, instead of reaching into taxpayers' pockets, would allow farmers to draw on more capital when times are tough. If we could get farmers to reduce their financial share in the farm, it might have some benefits to taxpayers in the future. In the past, public sympathy for farm relief came partly from pity for farm families on account of the huge financial losses they were suffering, but with farmers selling equity to doctors, there might be much less political support for subsidies. Farms might even be thought of, for once, as business ventures—which, after all, is what they are.

CONCLUSION

Although history shows that farm subsidies will not solve the "farm problem," there is still a role for government: It can provide a stable economic environment within which our farmers can produce and compete. When the Fed put on the brakes in the early 1930s, prices plummeted; farmers did not anticipate the massive price decline, and could not service a debt based on higher prices. Similarly, a major cause of the farm crisis of the 1980s was the inflationary monetary and fiscal policy of the 1970s. Inflationary expectations furnished part of the fuel for the escalation of land values up to 1980; crushing these expectations contributed to the recent years of agricultural distress. Must we continually repeat history? Perhaps not. If we can maintain a stable economic environment, we can surely lessen the chance of future booms and busts in American agriculture.

7

How Can Displaced Workers Find Better Jobs?

ELIZABETH HOFFMAN

Americans have long had a love–hate relationship with technology. On the one hand, we embrace the new consumer products made possible by technological change, while on the other we fear the loss of jobs. We want to have a growing economy and new products, but there seems to be some trade-off between economic growth and job security. Faced with the reality of technological unemployment, workers have tended to react first with militancy and a sense of outrage at having lost their jobs, even though there have generally been well-paying jobs requiring new skills available. The example, however, of textile production in early industrial Britain, and the tales of two American cities—Lowell, Massachusetts, and Pittsburgh, Pennsylvania—teach that workers have eventually learned new skills, and often higher incomes, as a result of technological change.

SMASHED LOOMS, HIGHER WAGES

Technological unemployment first became a social problem in the cotton textile industry in late-eighteenth and early nineteenth-century Britain. Before then, spinners and weavers had worked out of their homes on spinning wheels and hand-operated looms. Children could expect to, and generally did, follow their parents' employment. A hand-loom weaver could train his son to be a weaver and a hand spinner could train her daughter. In the 1770s, however, a series of power-driven mechanical inventions in cotton spinning, including what is called the *mule spinner,* multiplied the quantity of yarn that a single spinner could produce in a day by up to several hundred times. By the end of the decade, competition from machine-spun yarn made it impossible for a hand spinner to earn a living wage. There were growing demands and good wages available for workers willing to work the mechanical spinners.

The catch was that the new machines required water or steam power. Since a water mill or a steam boiler could produce power for several machines, it was more profitable to have a number of machines in one factory and bring the workers to it. Changing from a hand spinner to a machine spinner thus required more than just learning a new trade; it required a wholesale change in one's life.

As mechanical spinning machines were adopted, the price of cotton yarn fell, and output increased dramatically. Weaving, however, was not yet mechanized and there were not enough hand-loom weavers to keep up with the growing demand for their services. As weavers' wages rose, entrepreneurs and inventors began experimenting with weaving machines. The power loom was invented in 1787, and finally perfected in the 1820s; by the 1830s, two workers on a power loom in a factory could produce, in one day, twenty times what a hand-loom weaver could produce at home. The golden age of the hand-loom weaver was over, and by 1840 there were only a few left working.

These early displaced workers responded with a militancy far surpassing that of, say, steelworkers recently laid off in the United States. The displaced weavers, calling themselves Luddites, after a spinner named Ned Lud, burned factories and smashed machines. Their cause, however, was doomed. For every worker wanting to stop the pace of technological change, several others were happy to work for a factory's higher wages.

LOWELL: EXPERIMENTS IN CHANGE

The industrial history of Lowell, Massachusetts, shows that the effects of technological change on employment were as dramatic in the United States as in Britain. Built at Pawtucket Falls on the Merrimack River, the city had depended on lumber, grain, gunpowder, and cloth preparation until 1821, when a group of Boston investors founded the Merrimack Manufacturing Company and decided to purchase land to build a factory town. Utilization of this site depended on two crucial technologies: machines and the water power to drive them. Water power was supplied by a system of locks and canals that harnessed the power of Pawtucket Falls, but the machines were more difficult to come by.

At this time the British power loom was still a well-kept industrial secret, but Francis Cabot Lowell toured British mills and studied the machinery he saw. After he returned to the United States, he began tinkering with his own design. By 1817 he had perfected a working version of the loom, and is today credited with "reinventing" the loom in America.

In February 1822 the Merrimack Manufacturing Company placed its first order for machinery for spinning, weaving, and finishing cotton cloth. All that remained was to recruit and train a labor force to run the new machinery. The investors launched the "Lowell Experiment." They built factory dormitories and offered the impoverished farm girls of New England good pay, clean working conditions, and supervised living. The $2.25–4.00 per week, with a mandatory deduction for room and board, was *more than these women could earn in any other occupation open to women at the time.* The response was immediate. Women rushed to Lowell to work in the factories. Production began in September 1823, and the first shipment of cotton cloth left the factory on January 3, 1824. The city was incorporated on March 1, 1826, with a population of 2,500. Ten years later, the population had swelled to nearly 18,000.

Most of the population of Lowell consisted of young female mill workers who worked twelve to fourteen hours a day, six days a week, for an average of four years. They typically saved most of their net earnings for a dowry or to send a brother to school; their wages were sometimes their families' sole support. What little the girls did not save was spent on library subscriptions, concerts, and lecture series. At a time when young women were not allowed to attend college, the company arranged for Harvard University professors to come to Lowell to address them. These women even edited a literary magazine, the *Lowell Offering,* which became famous throughout the United States and Europe.

In the 1840s, perhaps because of competition from other New England mills, the mill owners tried to cut costs by instituting various incentives to speed up production. Piece rates were cut, making it necessary to work faster in order to make the same income. In addition, supervisors were given bonuses for greater output from their divisions. Workers who only had been expected to tend one or two looms before were now expected to tend three or four. Work that had been difficult at the earlier wages now became intolerable to many of these women. Some of the women responded with a series of strikes and efforts to persuade the state legislature to institute a mandatory ten-hour day. Others refused to speak out and continued working, perhaps fearing that a ten-hour day would just reduce the amount they could save and extend their stay at the mills. Still others simply went home or to the Midwest.

Changing labor supply conditions allowed the mill owners finally to win their battle with the mill workers. After the Irish famine of 1848, thousands of Irish families migrated to Lowell, willing to work at the wages offered and under the working conditions that existed at the time. Next came the French Canadians. After 1890 there were Greeks, Poles, Lithuanians, Italians, and Portuguese. Lowell had become an international melting pot.

Now, instead of primarily employing women, the mills employed whole families. This allowed the owners to take advantage of the greater diversity of skills that both sexes and all ages can bring to the workplace. Men could work the machines that required strength faster than could the women, while the women could perform work requiring manual dexterity faster than the men. The specialization of work by gender was certainly not complete, but the workplace was quite different than it was before.

The next set of technological changes came at the end of the nineteenth century, when the ring spindle drastically transformed the work requirements in spinning and weaving. The mule spinner and power loom, which had been used for many years, required some strength and skill to operate. A worker who tended the mule had to check for broken threads; when threads broke, he or she would have to shut down the machine, repair the breaks, and then restart the apparatus. By contrast, the ring spindle automatically shut down when threads broke and the automatic loom eliminated many of the manual steps needed to operate the older power loom. A mill using ring spindles and automatic looms could operate with unskilled labor and only a few skilled supervisors to take care of broken-down machines.

Attempts to substitute the new technology were fought by workers because they tended to upset the division of labor by gender. Work that had been performed largely by men after the farm girls left was now easily done by women. Since the women earned lower wages than did the men, it was often more profitable to hire women to run these new machines. Thus, men felt themselves technologically unemployed, while women willing to work on the new machines saw an increase in demand for their labor. The men worked as supervisors, but there were fewer supervisor jobs.

The adoption of the automatic loom and the ring spindle also facilitated a geographical change in textile production. Wages had been lower in the South than in the North at least since the Civil War, but skill levels were also lower. Under the older technology, which required more skill, the northern mills with their highly skilled and better paid labor force had a cost advantage over southern mills. The new technology, which demanded unskilled labor, allowed southern mills to take over the entire textile market by the 1830s.

With the movement of the textile industry from Lowell to the South, the northern mills were used to make carpets, shoes, sweaters, and various other industrial products. Despite the substitution of other means of employment, however, the Great Depression was a time of supreme hardship for the factory workers of Lowell. Prosperity only returned with mobilization for World War II and the growth of the electronics industry after the war. At that time Lowell was a place where labor was relatively inexpensive and there were men and women willing to work. By the 1970s, Lowell was in a slump once again: The domestic electronics and shoe industries had not kept up with foreign competition, and the factories were forced to reclose. In the 1980s, however, Lowell rebounded in response to the growth in services and high-tech industries.

The story of the ups and downs of Lowell, Massachusetts, also could be told of many industrial cities in the United States and Europe. Some are still great cities today, employing people to do tasks unheard of a century ago. Others are ghost towns to past technology.

THE PITTSBURGH MODEL

A more recent success story is Pittsburgh. Throughout most of its history, Pittsburgh's economy depended on steel making. There, as with textiles in Lowell, technological changes in steel destroyed and created jobs, but making steel was a constant of life. During most of the twentieth century, steelworkers learned their trades on the job and earned more than the median national income doing manual labor; however, during the 1970s the U.S. labor market changed markedly. Technological changes abroad, and relatively expensive labor at home, made it unprofitable to produce steel in the United States. Workers initially responded to unemployment by waiting for the industry to rebound and by lobbying Congress for tariffs and quotas to stem the import of foreign steel, but it eventually became obvious that Congress was not going to make foreign steel either unavailable or prohibitively expensive. The demise of the steel industry forced Pittsburgh to either change or face a fatal decline.

The city responded by mobilizing to create a new way of life. Strict pollution control laws were enacted and enforced, turning a city that only steelworkers could love into a beautiful place in which to live. Light manufacturing, service, and banking establishments relocated in Pittsburgh to take advantage of the labor force. The growth of high-tech industries led to an increase in opportunities for college graduates with training in engineering, computer, and health care skills. When a local television news show in Pittsburgh announced over a weekend that free tuition to the local community college was available for unemployed steelworkers, 13,000 former steelworkers enrolled that Monday morning. Many of these former steelworkers ended up with better paying jobs, and are now designing robots for the automobile industry or working as hospital technicians. By the early 1980s, Pittsburgh had become a model for other former steel towns; representatives from outmoded factory districts as far away as Germany and France came to Pittsburgh to study the way the city had successfully made the switch.

CONCLUSION

The history of technological change and worker displacement illustrates a fundamental economic principle that provides an important lesson for the future. In the United States, the marketplace tells workers which activities are currently profitable and which are not. When a major technological change is underway, the market signals to workers through higher relative wages the relative advantages and disadvantages of different occupations. With this information, workers can decide whether or not to seek training to use the new technology. When textiles were replaced with electronics, workers learned new skills once again.

   Similarly, when Pittsburgh was transformed from a steel town to a service
center, workers retrained. The other side of this story, however, is that
the market also tells workers which activities are no longer profitable. If
skills become obsolete, then demand for labor with those skills declines
and the wages of those workers decline relative to wages in other occupations. This process signals to workers that they should change jobs or
else suffer a loss in living standards. Hand spinners' and hand-loom weavers' wages declined when the mule spinner and the power loom became
profitable. Spinners and weavers either moved to the factories or worked
at totally different occupations. Steelworkers earned less in the late 1970s
and early 1980s than they had in the 1950s, and most were eventually laid
off. The older steelworkers retired early, the younger ones prepared themselves for new careers. Today we face a world in which technology is
changing more rapidly than ever before. While technological change has
been occurring since the beginning of society, it has become sufficiently
rapid so as to make radical changes in the structure of employment from
one generation to the next in the last 250 years. In the nineteenth century
each *generation* needed to learn different skills; the son of a hand-loom
weaver might become a power-loom operator or a steelworker. Today,
each generation may have to retrain *several times* in order to keep up with
technological change. Consequently, any purely vocational training is only
useful for a limited amount of time. The challenge for the future will be
to train each generation for a lifetime of change rather than for a specific
skill or job. This task suggests that the kind of education that will best
prepare the next generation is an education in flexibility: learning to learn
new things. Learning cannot simply stop at the end of school if workers
are going to continue to adapt to and profit from changing technology.

III
WOMEN AND MINORITIES

Who gets left out? Certainly slaves and their descendants; certainly women. We should do something about it. A problem is that the only "we" available for instant redress is the government, especially the one government we Americans have in common. Here are a couple of second thoughts, by Robert Margo and Elyce Rotella, on what government did to African-Americans and to women. "Did to" is sometimes the right verb-phrase, although it is always clothed as "did for."

Policy toward the powerless raises the First Question about policy: Should there be a policy in the first place? It is worth asking anytime, but especially when the people to be benefited are by assumption left out. The powers that be are to help the powerless. That is a trifle odd. Perhaps, as Margo and Rotella suggest, they will do better without the agent from the government, who is there to help them.

8

What Is the Key
to Black Progress?

ROBERT A. MARGO

In his landmark 1943 book *An American Dilemma,* Gunnar Myrdal described the economic conditions of black Americans as "pathological." Except for a small number "enjoying upper or middle class status," the "masses of American Negroes" were "destitute." He continued:

> They own little property; even their household goods are mostly inadequate and dilapidated. Their incomes are not only low but irregular. They thus live from day to day and have scant security for the future.

Although Myrdal's gloomy assessment held out little hope for the future, the per capita income of blacks, relative to whites, has risen substantially since World War II. A proximate cause has been an increase in the proportion of black men and women entering skilled and white-collar occupations. Approximately 53 percent of employed blacks held such jobs in 1988, compared with 9 percent in 1940. A full one-third of all black families in 1988 enjoyed the social and economic benefits of an income over $35,000. If one measure of success is a cover story in *Time* magazine, then the black middle class has clearly arrived.* What caused the emergence of a black middle class, after decades of apparently limited racial progress? Many would cite the Civil Rights Movement, and the antidiscrimination legislation it brought forth. Contrary to popular belief, however, the preconditions for a growing black middle class were in place before the Freedom Riders and the March on Washington. Middle-class blacks owe their successes in large part to themselves, and to the sacrifices of past generations of black parents who, in the face of great adversity, sent their children to school—and who, like European immigrants, moved great distances so that their children might have a better life. As is frequently done, to

*Richard Lacayo, "Between Two Worlds," *Time* (March 13, 1989), pp. 58–68.

interpret the growth of the black middle class, independently of the long-term trends in black schooling and migration, is to interpret it in a historical vacuum, and to miss something of significance for current public policy.

THE RACE–INCOME GAP

Economic historians have traced out the long-term evolution of racial income differences, and much more is known today than even a few years ago. At the end of the Civil War the vast majority of black Americans were recently freed slaves. They were mostly unskilled, impoverished, rural, agricultural, and Southern, whose annual incomes were perhaps no more than a quarter of those of whites. Recent research points to a narrowing in racial income differences by the end of the nineteenth century. Average black incomes may have risen to one-third of white incomes by 1900, and the rate at which Southern blacks accumulated wealth—in real estate and other forms—exceeded the rate at which Southern whites accumulated wealth. Relative improvement slowed, however, between 1900 and 1940. By the time America had entered World War II, the black–white income ratio was only slightly higher than it had been in 1900; all the gains took place before 1930, indicating that blacks suffered disproportionately from the hardships of the Great Depression. Nevertheless, between 1940 and 1980, the race–income ratio increased by twelve percentage points, rising four percentage points alone during the 1940s. A full half of the postwar increase occurred before 1960—*before* federal antidiscrimination legislation could have had much effect.

It thus appears that racial income differences were narrowing before the marches and sit-ins of the 1960s, although at an unsteady pace. Real income per person in America as a whole rose fivefold from 1870 to 1960, and the incomes of black people not only matched this dizzy growth—it surpassed it. This was a time, historians tell us, when black people could not vote, when racial segregation and discrimination in most every aspect of life was widespread and either legal or tacitly sanctioned. How, then, did black Americans manage to progress at all?

GAINING GROUND AT THE SCHOOLHOUSE

In 1890, the first year for which fairly reliable data are available, about 30 percent of Southern black children ages 5–20 attended some kind of school. Some things do not change—then, as now, skipping school was a good way to stay illiterate. According to the census of 1900, 37 percent of Southern blacks between the ages of 15 and 24 were *completely* (not just functionally) illiterate. By 1920, the proportion in school in the 5–20-year-old age group had increased to 53 percent and the illiteracy rate was down to

18 percent. By 1950, 69 percent were in school, the same as the white attendance rate. Over time, the quality of schools black children attended improved as well.* The improvement was greatest in the South, which started behind the North, and no change was more significant than the lengthening of the Southern school year: In 1950 the average length of the school year in the South's black schools was 173 days—below the Northern average, but 53 days more than it was in 1920. Each successive generation of black children attended school for more years than had previous generations. When they entered the labor market, they were better educated than were their parents.

Still, black children have faced serious obstacles in education. One of slavery's most pernicious legacies was adult illiteracy; perhaps 90–95 percent of newly freed adult slaves could not read or write. Children of illiterates attended school less frequently than did children of persons who could read, and if the local schools failed to teach them, what could their parents do to compensate? As literacy eventually spread through the black population, however, this particular consequence of slavery loosened its intergenerational grip.

Adult illiteracy was bad enough, but the schooling of black children was also hampered by educational discrimination. In *Plessy* v *Ferguson* (1896) the Supreme Court established the principle that de jure segregated public facilities, including schools, had to be "equal," but equality was hardly to be had, since the equal part of separate-but-equal was a myth. For example, for every dollar spent per pupil on instruction in Alabama's white schools in 1910 only 27 cents were spent per black pupil. The Court reversed the separate-but-equal decision in *Brown* v *Board of Education* in 1954, but the damage was done. Had the separate-but-equal decision been enforced throughout the first half of the twentieth century, then the long-term improvement in black people's schooling would have been greater. That the racial differences in schooling did eventually narrow is testament to the commitment of past generations of ordinary black men and women to educate their offspring, in spite of the obstacles put in their way. Thus, black parents are the unsung heroes of the Civil Rights Movement.†

---

*There is an unfortunate myth in our country that schools were once much "better" than they are today. We are so accustomed to negative images of big-city classrooms that the myth seems plausible. We are right to be concerned, but not because the good old days were so good. Consider the plight of a black child growing up in the rural South at the turn of the twentieth century: The schools were racially segregated, and typically might have stayed open three months per school year. It is likely that the state in which a student lived had no compulsory schooling law, or if it did, the law was barely enforced. If the student attended school at the average rate the total number of days spent in school between the ages 5 and 17 might equal roughly 250–300, not much more than a diligent child would attend today over the period of a year and a half. Today's urban teacher complains of overcrowded classrooms, inadequate supplies, and low pay, but who among them would rather teach in one of South Carolina's rural black schools circa 1920, with an average class size of 50 or more, in a one-room unheated building with no indoor plumbing and no textbooks? Simply put, our country's schools were much better on the eve of World War II than they were in 1890.
†For a detailed discussion of the points in this section, see Robert A. Margo, *Race and*

THE GREAT MIGRATION

Another way blacks raised their income was by leaving the rural South. Millions of black people left during the "Great Migration" from 1910 to 1960. The vast majority headed for a small number of major urban centers in the North and the West. Few made their way before World War I, in part because established black communities outside the South were still small in size. With the cutoff in immigration during the war, however, the North's manufacturing plants became hungry for labor, and black labor increased dramatically. The migration continued during the 1920s, though slowing to a halt during the labor-surplus years of the Great Depression. World War II again hastened the exodus, this time for good. In 1960, 41 percent of black Americans lived in the North, up from 10 percent at the turn of the century.

By leaving the South a black could substantially increase one's real income. The cost of living was higher in the North than it was in the South, but wages were higher still. Nor were the options unrelated. Moreover, contrary to popular opinion it was not the poor, uneducated, unskilled black farm laborer who was most likely to leave the South. Rather, the odds of migrating rose sharply with educational attainment, and with good reason: The vast majority of jobs filled by Southern black migrants were nonfarm jobs in which even a modicum of schooling was valuable. Black newspapers and letters from migrants carried reliable information about conditions up North back to the South—written news, meant to be read. Black migrants were ambitious, hard-working, earning more than Northern-born blacks with comparable experience. Their children, too, benefited from the move. De facto segregation was not any better than de jure segregation, but Northern schools were generally better in quality than Southern schools.

None of this is to say that schooling offered an easy ride to economic freedom. Occupations and industries were still highly segregated by race, in both the North and the South; it appears that employment segregation in the South was increasing between 1900 and 1950. Prior to the Civil Rights era, blacks could not aspire to high-paying white collar jobs in sales, management, or clerical or blue-collar supervisory positions. The number of black elected officials was negligible, government employment was severely circumscribed, and access to on-the-job training was jealously restricted by unions and white employees. History offers no evidence that racial economic differences could have been overcome solely by the long-term trends in black schooling and migration. The *demand* for black labor had to increase, and the record shows that such increases were coincident with large-scale "shocks" to society, such as the two world wars, and the Civil Rights Movement.

Schooling in the South, 1880–1950: An Economic History (Chicago: University of Chicago Press, 1990).

CONCLUSION

Let us recall events of the late 1950s, and imagine that the trends in schooling and migration had never taken place. What if the black labor force, poised on the eve of the Civil Rights Movement, was just as illiterate, impoverished, rural, and Southern as when Lincoln freed the slaves? Three decades later, would we have as large a black middle class as we do today? Plainly not. Three members of the United States Commission on Civil Rights say it well. "The purpose of outlawing employment discrimination," "wrote Mary Frances Berry, Francis Guess, and Blandina Cardenas Ramirez in 1986, "was to increase the opportunities for *qualified* [emphasis added] blacks to gain better paid employment."

History establishes that the link between labor market opportunities and qualifications was, and is, crucial for all Americans, black and white. The children of the black middle class will prosper, but beneath the black middle class lies the black underclass, increasingly undereducated and isolated from the mainstream economy. If two decades of social science research, most of it ahistorical, has taught us anything, it is that the problems of the black underclass defy quick fixes. Schooling, however, as obvious as it seems, is central to any fix. Without more and better schooling, underclass black children are no more likely to escape poverty and deprivation than the illiterate, postbellum Southern black could escape the rural South. The cost to the nation of ignoring this lesson may be the reversal of some of the hard-won economic gains of generations past and of the struggle for Civil Rights.

9

The Equal Rights Amendment—
Yes, But *Whose?*

ELYCE J. ROTELLA

Should the government establish and enforce laws that require women and men be treated differently? Is it fair for the government to give protections to women that it does not give to men? Can we justify laws that restrict women's right to do things that men are free to do? Should the U.S. Constitution be amended so that "equality of Rights under the law shall not be denied or abridged by the United States or by any State on account of sex?" Throughout U.S. history, many kinds of laws have required that women workers be treated differently from men. These laws have limited the hours that women work, forbidden the employment of women in certain occupations, required special facilities for female workers, and set minimum wages for women. Usually such laws have been motivated by concerns for women's health and well-being, and for this reason are commonly referred to as "protective labor laws." Often, however, these laws have had the effect of limiting women's opportunities in the labor force, and have become what legal historian Judith Baer has called "chains of protection."

UNEQUAL RIGHTS

Laws have governed some aspects of the contract between workers and employers since colonial days, but agitation for limits on hours of work began in the 1830s and 1840s, when textile mill workers unsuccessfully petitioned the Massachusetts legislature for legal limitation of the work day to ten hours. Although five states passed ten-hour day laws in the late 1840s, these proved to be unenforceable, and further attempts to regulate the hours of *all* workers were unsuccessful until the Fair Labor Standards Act was passed in 1938 and was upheld by the Supreme Court in 1941.

Attempts to regulate hours and other conditions of employment were renewed by progressive social reformers in the late nineteenth century. They first turned their attention to employed children, arguing that long hours and poor working conditions impaired children's health, and then used the same health arguments to push for legislation concerning women. As a result, Massachusetts passed the first enforceable law to restrict the working hours of adult women in 1879, setting the maximum factory work-day at ten hours and the maximum work week at sixty hours.

Many states followed Massachusetts' lead and passed protective legis-lation for women workers in the 1880s and 1890s. In some cases the state courts upheld the laws, but the laws were annulled in others. In 1895, the Illinois Supreme Court struck down an eight-hour day for factory women on the grounds that it interfered with the right of contract and that sex alone was not a justification for the act. By contrast, the Pennsylvania Supreme Court upheld a twelve-hour day law for women in 1900 and laid down the judicial philosophy of women being grouped by themselves as a distinct class in need of protections not needed by men. The Pennsylvania court wrote:

> Adult females are a class as distinct as minors, separated by natural conditions from all other laborers, and are so constituted as to be unable to endure physical exertion and exposure to the extent and degree that is not harmful to adult males.

This judgment drew parallels between women and children as distinct groups and viewed women, like children, as having special weaknesses that required protections not needed by men.

The most serious challenge to the new legislation came in 1908, when an employer, Mr. Muller, challenged the constitutionality of a 1903 Oregon ten-hour law for women on the grounds that it violated the "right to liberty" provided by the Fourteenth Amendment. Still, although the U.S. Supreme Court had previously struck down a New York hours law for *all workers* as "an unreasonable, unnecessary, and arbitrary interference with the right and liberty of the individual to contract in relation to his labor," the Court now ruled in favor of protective laws if applied only to women.

In their brief, Justices Louis Brandeis and Josephine Goldmark agreed that the right to purchase or sell labor was indeed a part of the liberty guaranteed by the Fourteenth Amendment, but they noted a previous Supreme Court decision that this right is "subject to such reasonable re-straint of action as the state may impose in the exercise of the police power for the protection of health, safety, morals, and the general welfare." They then went on to quote a host of doctors, scientific studies, books, judges, and other experts in an attempt to convince the court that women are physically weaker than men, that working long hours has a deleterious effect on women's health and morals, and that the general welfare, in addition to the welfare of individual women, would be undermined because future generations would be impaired if the health of mothers and potential mothers were weakened by long hours of work. In other words, women were not the physical equals of men, and therefore should not have equal rights to employment. These arguments proved to be persuasive, and the Court upheld the Oregon law establishing federal precedent for protective laws. Writing the opinion for the majority, Justice Brewer said:

> That woman's physical structure and the performance of maternal functions place her at a disadvantage in the struggle for subsistence is obvious . . . continuance for a long time on her feet at work, repeating this from day to day, tends to injurious effects upon the body, and as healthy mothers are essential to vigorous offspring, the physical well-being of woman becomes an object of public interest and care in order to preserve the strength and vigor of the race.

After the *Muller* decision, most states passed legislation to limit the work hours of women. By 1936, all but four states had hours restrictions that covered women's employment in most occupations except domestic service and farm work. Laws that prohibited night work and overtime work, forbade the employment of women in mines and some other occupations, required seats and rest periods, and set minimum wages for women were also popular. There were court challenges to these laws, but most were upheld on health-related arguments. It is interesting that proponents of minimum wages focused on morals rather than on physical health. They argued that low wages forced women into prostitution.

ORGANIZED LABOR VERSUS EQUAL RIGHTS

Legislation restricting a woman's right to work received strong support not only from progressive social reformers, who believed that such laws increased the well being of working class women, but also from trade unions—many of which excluded women from membership and were interested in protecting their members from competition by women workers. Union sentiment is well represented by an 1879 statement of the Cigar Makers Union that, after noting that the employment of women had increased in "alarming proportion," went on to say "[We] cannot drive the females out of the factories, but we can restrict their daily quota of labor through factory laws."

The American Federation of Labor (AFL), in particular, lobbied against equal employment rights for women. In 1892 and 1894, the AFL endorsed the eight-hour day for women and prohibition of female employment on foot-powered machinery. At the 1898 AFL national convention, a resolution asked Congress "to remove all women from government employment, and thereby to encourage their removal from everyday walks of life and relegate them to the home."

Before the 1920s, few female voices were raised against the restriction of a woman's right to work; support for equal employment opportunities came primarily from employers, who disliked government interference in their freedom to contract with workers. Some women did protest that such laws forbade them from taking jobs that they wanted or caused them to lose jobs; one study found evidence of women losing jobs in newspaper offices, foundries, and in transportation because of hours and nightwork restrictions. Some waitresses complained that nightwork restriction kept them from working during the hours when tips were best. Women who had taken jobs in streetcars during World War I, when hours restrictions were relaxed, protested their firing from these jobs when hours restrictions were reinstituted after the war. The voice of organized labor, however, was louder, or at any rate, more threatening. In Cleveland in 1918, the union representing male street car conductors struck to protest the employment of women and succeeded in getting the company to agree that "there will be no more women employed as conductors, and the Cleveland Railway Company will remove and displace the women that are now in service. . . . "

WOMEN'S RIGHTS: A RIGHT-WING MOVEMENT?

Organized opposition by women to protective legislation began in the 1920s when Alice Paul, leader of the newly formed National Women's Party, first proposed the Equal Rights Amendment. Prior to 1920, politically active women had been united in the struggle to win the vote. After the passage of the Nineteenth Amendment, however, deep divisions appeared within the ranks of the suffragists. Paul and her followers concentrated their energies on working for the passage of the Equal Rights Amendment, which Paul had written. Although the amendment would prohibit denial of any rights, including the right to employment, on the basis of sex, progressive social reformers, who comprised the majority of politically active women, saw the amendment as a threat to the workplace legislation they had worked so hard to pass. Paul acknowledged the threat of the ERA to such legislation, but argued that women were more hurt than helped by laws that restricted their opportunities for employment.

At the beginning, the National Women's Party was nearly alone in supporting the ERA. Almost all other women's groups—including the League of Women Voters, the American Association of University Women, the Women's Bureau of the Department of Labor, the Parents and Teachers Association, the Women's Christian Temperance Union, and the National Council of Jewish Women—were adamantly opposed. The dispute was so bitter that it has come to be called the "women's war."

On both sides were people who genuinely believed that their stance would lead to greater well being for women. Those who opposed the ERA believed women to be particularly subject to exploitation in the labor market and therefore in need of special protections. Those in favor of the

ERA argued that workplace legislation reduced women's freedom to operate in the labor market, thereby restricting their opportunities and lowering earnings.

In addition, there were groups on both sides whose motives were not related to concerns for the well being of women. Opposing the ERA were trade unions and the AFL, who feared competition from women workers. Favoring the amendment were employers groups who wanted the freedom to make decisions about hours, wages, and working conditions without government interference.

In marked contrast to the positions taken in the 1970s and 1980s, however, support for the ERA in the 1920s came overwhelmingly from the political right and opposition from the political left. In terms of partisan politics, Republicans favored the ERA, while Democrats led the opposition. How can we explain this dramatic change in the political orientation of ERA supporters and opponents? Why did the Republican Party and the political right abandon its traditional support of equal rights for women? Why did the Democratic Party and the left embrace a cause that it had so long fought?

INDIVIDUALISM OR COLLECTIVISM?

Throughout the long struggle for the ERA, and during earlier disputes about protectionist laws, those who favored special protections and restrictions were those who saw women as a *group,* who analyzed social problems in terms of groups and *classes,* and who therefore took a *collectivist* approach to reform. In the late nineteenth and early twentieth centuries these were Progressives, who worried about the welfare of the working class and saw women as an oppressed group who required group remedies. In the 1920s and 1930s, these were politically active ex-suffragists who believed that working women would suffer economic exploitation without the protection offered by the laws threatened by the ERA. During the struggle for ratification in the 1970s and 1980s, opposition to the ERA focused on a social rather than an economic agenda. These later ERA opponents, however, had in common with earlier opponents their view of women as a group distinct from men in ways that required the protection of special laws.

By contrast, ERA proponents in all periods have stressed the freedom of the *individual.* In the 1920s, they worried about the lost freedom and opportunities of women who were forbidden to work at night, or in foundries, or serving drinks from behind a bar, or in legally prohibited occupations. They worried about the unusually strong or able or ambitious women who had the qualifications needed to succeed in jobs that most women could not do. They argued that, as a result of protective legislation, women were crowded into a restricted set of occupations in which their increased supply lowered earnings. During the struggle for ratification, proponents asked that individual women be given the opportunity to compete on an equal basis with men in all areas of social and economic life. They asked that there no longer be two classes of citizens; that women be admitted to full adult citizenship with the same rights and responsibilities as men. The economist and philosopher, John Stuart Mill, writing in 1868 gave what is perhaps the most eloquent statement of the logic behind the individualist position:

> One thing we may be sure of—that what is contrary to women's nature to do, they never will be made to do by giving their nature free reign. The anxiety of mankind to interfere in behalf of nature, for fear lest nature should not succeed in effecting its purpose, is an altogether unnecessary solicitude. What women by nature cannot do, it is quite superfluous to forbid them from doing. What they can do, but not so well as men who are their competitors, competition suffices to exclude them from; . . . If women have a natural inclination for some things than for others, there is no need of laws or social inculcation to make the majority of them do the former in preference to the latter.

CONCLUSION

The question of whether the laws of the United States should be sex-neutral continues to be asked. The Supreme Court recently has tackled the legality of employment restrictions based on the effect of work environments on pregnant and potentially pregnant workers. Armed services regulations mandated by congressional legislation exclude women from combat-related posts, but the exclusions are being challenged, particularly after the roles played by some women in the invasion of Panama and in the Persian Gulf War. Many laws requiring differential treatment remain on the books. The Equal Rights Amendment already has been reintroduced in Congress. The question that lies behind these issues is whether it is appropriate to use the police power of the state to enforce differential treatment of people because of their sex. Where one stands on the issue is often determined by what one takes to be natural or divinely ordained. Those who believe that there is a fundamental difference between the sexes, one that goes far beyond the obvious physical differences to determine behavioral and social roles, often favor special protections for women and restrictions on women's behavior. Those who oppose legally mandated differential treatment of the sexes believe that there are no differences between the sexes so profound as to require legal intervention that would reduce the individual's right to self-determination.

IV
GOVERNMENT AND
THE ECONOMY

Americans have quarreled about big or small government since the Federalist Papers and the Whiskey Rebellion. The sphere in which everyone agreed the federal government should be ruler was foreign economic affairs, namely, trade and the value of the dollar. Hence the essays by Barry Eichengreen on the gold standard and Mark Thomas on trade. The federal lands could hardly be anything else but federal, without causing quarrels among the original states (Virginia claimed land between its northern and southern parallels out to the Mississippi). Hence Terry Anderson and Peter Hill's essay on land policy. The income tax, however, on which the expansion of federal power in the twentieth century was based, was by no means implied by the Founders. An amendment was necessary to the Constitution. Ben Baack and Edward Ray point out that military exigencies drove us to the tax, as they had in Great Britain. William Cobbett had called Britain's desperate experiment with income taxation during the Napoleanic War "that damn'd inquisitorial tax."

What limits, then, were there to an expanding government in the twentieth century? Very few, but democracy drove it, not some conspiracy of pointy-headed bureaucrats. The people demanded that the federal government take responsibility for the business cycle, with the result that in 1914 we came for the first time to have a central bank. John Wallis does not think it entirely accidental that we then had a Great Depression.

The limits to government in the economy were more the incompetence of the government to impose its will than an ideology of laissez faire. Americans like to be left alone, but they also like to interfere with each other's consumption of booze and payments of charity. Jonathan Hughes brings the dismal news that the Nanny State, to use the British locution, has always been popular with Americans.

10

As Good as Gold—
By What Standard?

BARRY EICHENGREEN

The gold standard is a monetary arrangement in which countries peg the price of gold and impose no restrictions on imports and exports of the precious metal. It is conventionally portrayed as the normal state of international economic affairs prior to World War I. The gold standard's subsequent disintegration has been cited as a cause of inflation, boom–bust business cycles, fiscal irresponsibility, and the instability of monetary exchange. In reality, however, the classic gold standard was a short-lived and peculiar institutional arrangement that did *not* moderate business cycle fluctuations, guarantee price or exchange-rate stability, or prevent recession and financial crises. Also, far from being the automatic monetary thermostat of textbooks, the gold standard was in reality a *delicate* mechanism whose stability required painstaking management by central banks. Today, with the countries of the European Community moving to fix their exchange rates once and for all, and the United States and Japan pondering international monetary reform, it is critically important that policymakers possess an accurate picture of how the gold standard worked.

SILVER AND GOLD

Great Britain remained on the gold standard almost without interruption for more than two centuries after 1717, with the notable exception occurring during the French Wars at the beginning of the nineteenth century. Britain, however, went onto the gold standard not as a matter of policy but because Sir Isaac Newton, as Master of the Mint, set too high a price on silver in 1717, inadvertently driving all newly minted coins in Britain out of circulation and transferring the nation from bimetalism, the backing of currency with both gold and silver, to a de facto gold standard. By contrast, other countries adopted the gold standard only in the final decades of the nineteenth century.

For most of that century, the United States and much of Europe clung to the fiction of bimetalism. Gold and silver coins were supposed to circulate side by side, as the authorities pegged the domestic currency price of each. Shifts in the supplies of silver and gold, however, defeated the attempts of governments to peg the prices of the two metals. Those committed to backing paper currency with precious metal were forced to choose between the two alternatives. Three factors swung the balance in favor of gold.

First, the development of the steam engine reduced the risks of counterfeiting. The smallest practical gold coins were still too valuable for use in everyday transactions. The solution was to introduce token coins valued at more than their metallic content. That the government stood ready to redeem those coins at more than their intrinsic value provided an irresistible incentive for counterfeiting. Minting these coins to high standards with steam-powered machinery greatly increased the costs of doing so.

Second, with the British pound already pegged to gold, other countries could insure the stability of their exchange rates against sterling by emulating the British example. This appealed especially to countries that traded with Britain and sought to borrow in London, significant considerations in an era when Britain was the world's leading commercial and financial power.

Third, and perhaps most important in encouraging the movement toward gold, was the series of massive silver discoveries throughout the nineteenth century. As large quantities of silver came on the world market, bimetalist countries committed to purchasing it at a fixed domestic-currency price were faced with the specter of runaway inflation. As the market price of silver fell relative to the official price, customers queued up at the doors of bimetallic-country central banks, clamoring to swap their silver for the bank's gold. Rather than permit their gold reserves to be depleted and subject themselves to silver inflation, bimetallic countries suspended silver convertibility and turned to gold.

Movement toward a truly international gold standard gained momentum after 1870. Germany used the indemnity received in 1871–1873 as a result of her victory in the Franco–Prussian War to purchase gold and establish a new gold-based currency unit. The Netherlands, Denmark, Norway, and Sweden suspended silver coinage and turned to gold, an example quickly emulated by France and her partners in the Latin Monetary Union (Belgium, Switzerland, Italy, and Greece). In 1879 the gold standard conquered the Atlantic, as the United States ended the Greenback period, when the price of the dollar had been allowed to fluctuate against the British pound. The gold standard reached into Asia as Russia and Japan embraced gold. Despite exceptions such as the Austro–Hungarian Empire, which never officially adopted gold convertibility, and the countries of Latin America, which repeatedly suspended it and allowed their exchange rates to fluctuate, a truly international gold standard had emerged by 1880. Still, this "golden age" was not the normal state of affairs before the twentieth century. It prevailed for no more than a third of a century prior to 1913.

THE GOLD STANDARD AND STABILITY

A central merit of the gold standard system, according to its proponents, is that it serves to guarantee price stability, tying the hands of governments otherwise unable to resist the lure of inflationary finance. The point has been made with admirable clarity by Alan Greenspan, current chairman of the board of governors of the Federal Reserve System:

> In the absence of a gold standard, there is no way to protect savings from confiscation through inflation. . . . The financial policy of the welfare state requires that there be no way for the owners of wealth to protect themselves. This is the shabby secret of the welfare statists' tirades against gold. . . . Gold stands in the way of this insidious process. It stands as a protector of property rights.*

To say that the gold standard is "moral," however, is not necessarily to say that it is as practical as its adherents would have us believe. True enough, the classic gold-standard era was one of price stability when viewed through the prism of the 1970s. That prism, however, distorts the picture as it appeared to people during the years when the gold standard prevailed. While the price level in gold-standard countries was little different in 1913 than it had been in 1873, prices fell steadily over the first two decades of that period at a rate of about one per annum, before rising over the two succeeding decades at the same pace. The American Populist Party agitated for silver coinage to halt the decline in commodity prices and to relieve the capital markets from financial stringency. William Jennings Bryan spoke for disaffected farmers when he complained that the economy was being crucified on "a cross of gold."

The mechanism whereby the gold standard was supposed to stabilize prices was the response of the gold mining industry. Economic growth put downward pressure on the price level, since as growth proceeded more goods chased the same amount of money. With the prices of other commodities failing, but with central banks continuing to peg the nominal price of gold, the relative price or purchasing power of yellow metal rose proportionately. According to proponents of the gold standard, prospecting was encouraged, inactive gold mines were reopened, and development of new techniques for extracting known gold deposits was stimulated.

In fact, there is little evidence of the gold-mining mechanism operating in response to monetary trends during the classic gold-standard era. Nineteenth-century gold mining was driven by a sequence of great gold discoveries occurring at intervals of roughly twenty-five years, most often

*Cited in *The Economist,* 26 September 1987.

unrelated to deflationary pressures or monetary trends. These gold discoveries resulted form the expansion of agriculture and ranching into
sparsely settled regions. For example, when James W. Marshall discovered
gold while building a sawmill for John Sutter on the South Fork of the
American River in 1848, sparking the great California gold rush, he was
not in the least motivated by changes in the real price of gold. Deflation
did encourage some reopening of mines and some small-scale prospecting,
but of all developments affecting the gold-mining industry, only the cyanide
process developed in the 1880s was stimulated directly by the real price of
gold. That it took twenty years, from the 1870s to the 1890s, for trends in
gold mining to reverse an ongoing deflation is evidence that this mechanism
did not work swiftly or powerfully.

Yet perhaps the classic gold standard was more successful in moderating
business-cycle fluctuations? If one's standard of comparison is the post–
World War II period, the answer is again *no*. Long-accepted estimates of
nineteenth-century GNP and unemployment indicate that fluctuations in
economic activity were considerably more pronounced in the gold-standard
period than they were between 1945 and 1973. Although recent research
suggests that these accepted estimates may overstate the volatility of the
macroeconomy prior to 1913, not even the revised estimates for the pre–
1913 period suggest that the gold-standard era was any more stable than
the fiat money regime that prevailed after 1945. Also, notwithstanding the
sporadic bank failures and increasing turbulence that have characterized
financial markets in recent years, today's instability pales in comparison
with the prevalence of bank failures and financial crises in the half-century
ending in 1913. One might argue that these crises might have been even
more severe in the absence of the gold standard. It remains the case,
however, that the gold standard was far from a guarantor of financial and
economic stability.

FOREIGN EXCHANGE VERSUS
DOMESTIC PROSPERITY

If the gold standard did not guarantee price stability or prevent recession
and financial crisis, can it nevertheless be defended as a bulwark against
exchange-rate instability? Even this is a narrowly "Eurocentric" view. Exchange-rate stability did not extend to silver-standard countries such as
China and the Central American republics, whose currencies fluctuated
against those of the gold-standard world. Countries on the periphery of
northwestern Europe, such as Portugal, Italy, and Bulgaria, joined the
gold standard, but, finding themselves unable to defend their fixed parities,
were forced to alter their exchange rates. This inability of less-developed
countries to defend their fixed exchange rates is most evident in the cases
of Argentina, Brazil, Chile, and Mexico, each of which maintained flexible
exchange rates for extended periods of time.

Yet in northwestern Europe, at the center of the gold-standard system, thirty years of exchange-rate stability prevailed. It endured in the face of remarkably large trade surpluses and deficits. The surpluses and deficits of the industrial countries of Europe were an order of magnitude larger than the ones in the 1980s, when trade imbalances were widely blamed for destabilizing exchange rates. Capital flows between nations reached levels never subsequently matched. Yet none of these disturbances to the balance of payments succeeded in destabilizing exchange rates in industrial Europe. What can explain this singular success?

The success of the gold standard at the center of the international system rested on two elements. The gold standard's credibility—the confidence invested by the public in the government's stated commitment to its maintenance—derived from the priority attached by the governments to gold convertibility. In the core countries—Britain, France, and Germany—there was little doubt that the authorities ultimately would take whatever steps were required to defend the central bank's gold reserves and maintain the convertibility of the currency into gold at a fixed price. If one of these central banks lost gold reserves and its exchange rate weakened, funds would flow in from abroad in anticipation of the measures the authorities would eventually adopt in order to stem reserve losses and strengthen the exchange rate. The exchange rate consequently strengthened of its own accord. Stabilizing capital flows thereby minimized the need for actual government intervention.

What rendered the commitment to gold credible? In part, there was little perception that policies required for external balance were inconsistent with domestic prosperity. There was little awareness that defense of the gold standard and the reduction of unemployment might be at odds. Unemployment emerged as a coherent social and economic problem only around the turn of the century. Before then, unemployment was ascribed to individual failings. There was little awareness that currency fluctuations could affect employment prospects.

Even where observers connected unemployment to the state of business, there was little tendency to relate the latter to interest rates or monetary conditions. Contemporaries had limited appreciation of how central-bank policy affected the economy. There was no well-articulated theory of how supplies of money and credit could be manipulated to stabilize production or reduce joblessness, like the theories developed by the British economist John Maynard Keynes after World War I.

The working classes, possessing limited political power, were unable to challenge this state of affairs. In many countries, the extent of the franchise was still limited. Labor parties, where they existed, rarely exercised significant influence. Those who might have objected that restrictive monetary policy created unemployment were in no position to influence its formulation. Domestic political pressures did not undermine the credibility of the commitment to gold.

Nor was there a belief that budget deficits or changes in the level of public spending could be used to stabilize the economy. Since governments followed a balanced budget rule, changes in revenues dictated changes in the levels of public spending. Countries rarely found themselves confronted with the need to eliminate large budget deficits in order to stem gold outflows. There existed firmly established norms concerning the distribution of the fiscal burden. For revenues, central governments relied primarily on import duties. At the prevailing stage of economic development, taxes on income or domestic activity were costly to collect. The individuals upon whom the burden of import duties fell, often purchasers of imported foodstuffs and other consumer goods, tended to be wage earners with relatively little political say. When revenue needs fluctuated, import duties could be adjusted accordingly. The need to eliminate a budget deficit did not automatically open up a contentious debate over the incidence of taxation. Governments could credibly promise to direct fiscal as well as monetary instruments toward balance-of-payments targets.

Thus, a particular constellation of political power, reinforced by prevailing political institutions, and a particular view of the operation of the economy provided the foundation for the classical gold standard system. This combination of factors—political institutions and influence on the one hand, the prevailing conceptual framework on the other—was the immediate basis for the credibility of the system.

THE INTERNATIONAL ORCHESTRA

Ultimately, however, the credibility of the prewar gold standard rested on international cooperation. When stabilizing speculation and domestic intervention proved incapable of accommodating a disturbance, the system was stabilized through cooperation among governments and central banks. Minor problems could be dispatched by tacit cooperation, generally without open communication among the parties involved. When global credit conditions were overly restrictive and a loosening was required, for example, the requisite adjustment had to be undertaken simultaneously by several central banks. Unilateral action was risky; if one central bank reduced its discount rate—the rate it charged its institutional customers—but others failed to follow, then that bank would suffer reserve losses as capital flowed out and might be forced to reverse course. Under such circumstances, the most prominent central bank, the Bank of England, signaled the need for coordinated action. When it lowered its discount rate, other central banks typically responded in kind. In effect, the Bank of England provided a focal point for the harmonization of monetary policies. It was, to use Keynes' famous phrase, the "conductor of the international orchestra." By following their conductor, the central banks of other countries coordinated the necessary adjustments.

In contrast, major crises typically required different responses of different countries. The country losing gold and threatened by a convertibility crisis had to raise interest rates to attract funds from abroad; other countries had to loosen domestic credit conditions to make funds available to the central bank experiencing difficulties. The follow-the-leader approach did not suffice, especially when it was the leader, the Bank of England, whose reserves were under attack. Such crises were contained rather through overt, conscious cooperation among central banks and governments. Other central banks and governments discounted bills on behalf of the weak-currency country, purchased its securities or lent gold to its central bank. Consequently, the resources upon which any one country could draw when its gold standard was under attack far exceeded its gold reserves. They encompassed the resources of the gold-standard countries as well. This provided countries additional ammunition with which to defend their gold parities.

Thus, in 1890, when the Bank of England suffered gold losses due to the Baring Crisis—the insolvency of the House of Baring, a leading British financial firm—it obtained a loan of £2 million of gold from the Bank of France. It secured an additional £1.5 million in gold from the Russian government. The mere announcement that these funds had been made available proved sufficient to stem the gold drain from the Bank of England. There was no need for much of the French gold to be ferried across the Channel. Anticipating that concerted international action to defend sterling would be forthcoming, speculators reversed the direction of capital flows so as to render that action unnecessary.

Again in 1906–1907, the Bank of England received assistance from the Bank of France and its other foreign counterparts. On other occasions the favor was returned: In 1898 it was the turn of the German banks and the German Reichsbank to obtain assistance from the Bank of England and the Bank of France. The smaller gold-standard countries of Europe—Belgium, Finland, Norway, and Sweden among them—repeatedly borrowed reserves from foreign central banks and governments.

Thus, what rendered the commitment to the gold standard credible was that the commitment to its maintenance was international, not merely national. That commitment was activated by international cooperation.

THE GOLDEN CRACK-UP

Even before World War I, the foundations of the classic gold standard were being undermined. The priority attached to the defense of the gold standard was no longer beyond question as increasing importance was attached to other, potentially incompatible policy goals. Extension of the franchise, the rise of political parties dominated by the working classes, and the growing attention paid to unemployment all suggested that a time might come when there would be a conflict between defense of the gold standard and other objectives.

Simultaneously, the traditional basis for international cooperation was growing increasingly tenuous. The quarter-century after 1871 had been distinguished by relatively few political and military conflicts among the Western European powers. The spread, however, of international political tension after the turn of the century undermined the leading European countries' readiness to cooperate. Equally important, the United States had not been party to the cooperative arrangements supporting the international gold standard. The absence of a U.S. central bank precluded American participation in these cooperative ventures. As long as the United States was not the leading user of gold reserves, as she became after the turn of the century, her failure to participate in these cooperative arrangements and the destabilizing impulses she imparted to the operation of the international system did not threaten the entire edifice. By the first decade of the twentieth century, however, the United States had grown too large and too influential to remain on the fringes. The traditional basis for international cooperation no longer sufficed. One rationale for creating the Federal Reserve System in December 1913 was to more effectively manage the American gold standard. The existence of a U.S. central bank might have provided a superior basis for the requisite cooperation. Unfortunately, the newly created Federal Reserve System proved insular and unappreciative of the advantages of international cooperation.

The political and economic preconditions for the smooth operation of the classic gold standard were already in decline at the turn of the century. World War I accelerated the trend. It extended the political sphere to encompass economic issues that had previously remained outside it. The credibility of the commitment to gold was challenged by an array of political and economic changes that shattered the particular constellation of political power on which policy had been predicated. To secure labor peace, wartime governments encouraged the spread of unionism, and the determination of wages and employment suddenly became political issues. Extension of the franchise and the growth of political parties dominated by the working classes intensified pressure to adapt policy toward employment targets. When employment and balance-of-payments goals clashed, it was no longer clear which would dominate. Doubt was cast over the credibility of the commitment to gold.

With the erosion of credibility, international cooperation became even more important. Yet it was not forthcoming. Three obstacles blocked the way: domestic political constraints, international political disputes, and incompatible conceptual frameworks. Interest groups with the most to lose were able to stave off adjustments in domestic policy that would have facilitated international cooperation. The international dispute over war debts and German reparations hung like a dark cloud over all international negotiations. The competing conceptual frameworks employed in different countries prevented policymakers from reaching a common understanding of their economic problem, not to mention agreeing on a solution.

Given this decline in credibility and cooperation, it is no surprise that the interwar gold standard, reestablished by the second half of the 1920s, functioned less smoothly than did its prewar predecessor. Due to the lack of credibility, fixed domestic-currency gold prices were subjected to early and repeated attack. Due to the lack of cooperation, central banks' defenses proved inadequate. By the end of 1931, the majority of countries were back off the gold standard for good.

CONCLUSION

Policymakers contemplating international monetary reform look back nostalgically on the classic gold standard. Yet the gold standard was neither the normal way of organizing international monetary affairs prior to 1913, nor was it a guarantor of price and output stability. The gold standard prevailed as an international monetary system for barely a third of a century before World War I. Prices were far from stable, and the record of output stability compares unfavorably with recent years. Admittedly, the gold standard succeeded in stabilizing exchange rates in the face of trade deficits and international capital movements, but only at the international monetary system's European center. That the gold standard functioned smoothly only in Europe and only for three or four decades prior to 1913 suggests that the preconditions necessary for successful operation were specific to that time and place. The credibility of the commitment to gold convertibility, upon which exchange-rate stability hinged, depended on a particular constellation of political and economic forces: a limited franchise, limited recognition of unemployment, limited awareness of the connections between monetary policy and state of the domestic economy. With these preconditions no longer present, a smoothly functioning gold standard would be impossible to reestablish today.

11

Who's Afraid of the Big Bad Trade Deficit?

MARK THOMAS

America's foreign trade deficit is commonly attributed to a fundamental decline in our ability to compete with foreign goods, both at home and abroad. The popular view is that America must wrestle back the initiative in world markets, and become "Number One" yet again. The dry stuff of the trade statistics are thus clothed with an emotional mystique—of America losing the world trade game, of losing national face, and of becoming weak in the world (losing out to Japan and Germany, against whom we fought and won a World War). Such rhetoric is at best misplaced, and may even be downright dangerous.

The demand for a "favorable" balance of trade is a return to the language of mercantilism, in which economic and political power was identified with inflows of bullion made possible by having more exports than imports. The world economy, however, is no longer ruled by gold, even if mercantilist sympathies endure. The idea that surpluses are inherently "good" and deficits inherently "bad" is a myth; the renewed call for trade legislation to correct the deficit by eliminating unfair trading practices and by mandating automatic trade restrictions in the event of "large" bilateral surpluses constitutes a misbegotten policy, with the lessons of the past not learned well enough.

REVOLUTION AND RAILROADMANIA

Trade deficits are nothing new to the American economy. Before 1880, they were the rule rather than the exception. This was especially true during the colonial period, when the scale of the deficit was at times even higher than in the worst year of the current "crisis": the deficit was more than 9 percent of national income in 1771, compared with less than 3.5 percent in 1987. Moreover, the contrast is not an aberration of a single, unrepresentative year. The trade deficit for the entire period between 1700 and 1775 averages out at a figure close to 4 percent of colonial national incomes, compared with a little over 2 percent for the 1980s.*

The pattern continued after independence. Over the entire nineteenth century, there were only twenty years when the United States did not run a deficit. The size of the shortfall fluctuated considerably over time, with the largest deficits occurring in the 1830s, the 1850s, the decade after the Civil War, and the 1880s. The timing of these swings was no accident. The 1830s saw the heyday of canal construction in America; the 1850s witnessed the first railroad boom, which was resuscitated in 1868 after the national tragedy of civil war; and the 1880s were the years of transcontinental railroad mania. The current account deficits—the excess of imported goods and services over exports—exactly mirrored the booms and slumps in transportation investment, and with good reason. As much as 60 percent of the money invested in canals between 1834 and 1844 came from foreign sources. Between 1865 and 1893 foreigners provided more than one-third of the financing for railroad track, locomotives, and rolling stock, buying up half of the total volume of bonds and a quarter of all the stocks issued by railroad companies.

Foreign interests were so eager to invest in America because it paid. The capital demands of the burgeoning American economy were too massive for the immature domestic financial intermediaries of the day. The relative scarcity of funds promoted high rates of return for investors; they were higher than in the more advanced financial markets of Britain and Europe. The result was a substantial inflow of foreign capital into the United States—coincident with a long-run deficit of goods and services.

A MOST VALUABLE LUXURY

The moral of this tale is straightforward, but too often it has eluded the financial and economic journalism of the present "crisis": the current account is only one part of the balance of payments, and not necessarily the most important. According to statistical convention, the current and capital sides of the international ledger must, by definition, balance each other out. Whenever we buy more goods and services from abroad than foreigners buy from us, we must give them something in return. The quid pro quo is assets. That is why the Japanese are buying up American real estate, from Rockefeller Plaza to Universal Studios. It is a way of settling up the

---

*The colonial trade deficit, however, was, counterbalanced by net exports of services to the rest of the world. Americans dominated the trans-Atlantic carrying trade; so much so that shipping earnings (a service income) were larger than receipts from tobacco exports. All in all, colonial America probably ran at most small deficits on the current account—the balance of earnings and payments from exchange of goods and services. We can be less sanguine about the present situation—even after accounting for net exports of services, the United States in the 1980s still ran significant deficits on the current account.

accounts, of compensating for the large deficit we have with them. As history so eloquently shows us, however, the statistical balancing out of deficits and capital inflows need not result from "problems" in the current account, caused by lack of competitiveness or whatever. In the nineteenth century, especially after the cotton boom of the 1830s, it was the current account that went into the red in order to balance out the heavy inflow of funds to finance American enterprise. The United States had more profitable investment opportunities than it had domestic savings to finance them. The British, Germans, Dutch, and French stepped in and made themselves (and our American forebears) richer. By 1914, Americans were in debt to the rest of the world to the tune of $7.2 billion (almost $125 billion at 1991 prices). The United States was, then as now, the largest debtor in the world. Did this make America a weaker economy, a more vulnerable society than otherwise?

The answer is a categorical, *No*. Indeed, if anything, it was foreign capital that gave Americans the opportunity to forge a strong, dynamic economy. This was recognized as early as 1791 when Alexander Hamilton, one of the greatest economic minds produced by this country, observed that

> Instead of being viewed as a rival [foreign capital] ought to be considered as a most valuable luxury, conducing to put in motion a greater quantity of productive labor and a greater proportion of useful enterprises, than could exist without it.

A decade earlier, Robert Morris, financier supreme to the new Republic, had said much the same thing: "Money lent by the City of Amsterdam to clear the forests of America would be beneficial to both." This was not mere Federalist rhetoric. Hamilton and Morris were describing the reality of colonial economic growth. It has been estimated that the accumulated debt of the colonies to Britain alone amounted to $40 million (about $1 billion at today's prices) by the time of the Revolution. Much of that was short-term credit, rather than long-term investment. Nevertheless, it was vital to the smooth working of the colonial economy. The plantations of the Chesapeake, the factor houses of Norfolk and Charleston, the trading concerns of New York and Boston, the iron works of rural Pennsylvania and the shipyards of Newport were all dependent on British finance. Immediately after the Revolution, John Lord Sheffield declared that "the greater part of the colony commerce was carried on by means of British capitals" with more than four-fifths of European imports being "at all times made upon credit." Foreign money oiled the wheels of colonial commerce as well as colonial prosperity.

THAT COMMANDING POSITION

Independence saw no weaning of the American economy from the mother country as far as investment was concerned. In fact, American dependence on British investors became so ingrained that when the market for American securities slumped in London in 1839, work on capital projects was checked throughout the eastern seaboard, and stopped entirely along much of the frontier. Canal builders and railroad barons, symbols and architects of American expansion, went directly to London for capital. Foreign investors were absolutely crucial to the success of the canal boom of the 1830s, from the Erie Canal on down. Even the railroads glorified by that most American of board games, Monopoly, were owned in large part by the British, Dutch, and French in the 1850s. Indeed, so much of the stock of the Philadelphia and Reading was in British hands in 1857, that when the board of directors searched for a new president, they went to London to find him. Foreign capital was no less crucial to the opening up of the transcontinental system. When Jay Cooke, a father of the American capital market, wanted money to expand his Northern Pacific Railroad in the early 1870s to connect the east and west coasts, he turned to Europe, going from one banking house to another to raise the $50 million he needed. The Crash of 1873 ruined Cooke and his scheme; however, when plans were unveiled twenty years later to reorganize the old Northern Pacific, so much of the stock was owned in Germany that the Deutsche Bank had to be included in the new syndicate. The Illinois Central, the Baltimore and Ohio, the New York and Erie, the New York Central—the list of pioneer American railroads dependent on European finance is impressive. No wonder that it has been argued that foreign capital "meant the difference between slow, self-generated development and the very rapid development" that did in fact occur.

Railroads were the most visible beneficiaries of foreign capital, but Europeans invested in every conceivable aspect of the American economy where profits could be made: oil in Wyoming, textiles in Massachusetts and Georgia, land and agriculture in Colorado and Kansas, fruit and oil in California, mining in Virginia, California, and Nevada. Money flowed from the London Stock Exchange, and from the bourses of Paris, Amsterdam, Berlin, and Vienna. Firms as crucial to the American Industrial Revolution as Western Union, AT&T, Eastman Kodak, Carnegie Steel, and General Electric, all borrowed from overseas large volume. Borrowing from abroad was very much part of the American way of doing business in the nineteenth century.

And, just as the likes of Honda and Toyota built plants on American soil in the 1980s, so the British, French and Germans built factories here a century ago. In a society perhaps more finely tuned to nativist sentiments, the foreign invasion was made less obvious. The Iowa Land Company, the Alabama Coal, Iron, Land and Colonization Company, the Texas Land and Mortgage Company, the Prairie Cattle Company, and the American Thread Company may have sounded quintessentially American, but they were in fact businesses set up and run by foreigners.

By the late 1870s, foreign capital had largely worked its magic. The inflow of funds, by underwriting internal improvements and industrial expansion, had transformed the United States from a capital-scarce to a capital-rich economy. Americans became richer and financial markets matured, creating a domestic pool of savings more equal to its investment tasks. As supply began to catch up to demand, the rates of return on American stock fell relative to overseas markets, and the United States became a less attractive repository of European money. As the dependence of American entrepreneurs on foreign money came to an end, the balance of payments entered a new stage; after 1874, trade deficits all but disappeared.

Indeed, by 1896 the tide had completely turned, for the United States itself was now beginning to lend and invest overseas. Much of that investment mirrored earlier experience, as American investors bought up stock in foreign railroads and factories. Some of it, however, was radically new, as domestic corporations became multinational, building and running factories overseas, instead of just backing them. The United States was becoming a significant presence in the global economy, reflecting both an abundance of capital and a comparative advantage in managerial and organizational skills and other human talents. Theodore Roosevelt could announce with good reason in 1901 that "America has only just begun to assume that commanding position in the international business world which we believe will be more and more hers."

A BEDEVILING MYTH

The volume of capital outflows after the mid–1890s was so large that it took only twenty years for the United States to eradicate its almost three-century history as a net debtor. This remarkable nature of the transformation was not lost on contemporaries. John Hay, who as a young man had been Abraham Lincoln's private secretary and was secretary of state by 1902, observed in a paean to another Republican martyr, William McKinley, that

> [T]he debtor nation has become the chief creditor nation. The financial center of the world, which required thousands of years to journey from the Euphrates to the Thames and the Seine, seems passing to the Hudson between daybreak and dark.

For many commentators since, the use of American capital to sponsor economic development overseas—whether through private enterprise or in public form as the Marshall or Lend-Lease Plans—was part of the country's newfound destiny as the heartland of freedom and progress. The ideal was so entrenched after a near-century of realization that its reversal, at least as interpreted in the popular press, came as a dramatic blow to the national ego.

In every year between 1896 and 1972, the current account was in surplus.

This period saw the birth of a myth that continues to bedevil discussion of the trade balance: the idea that surpluses are inherently "good" and deficits inherently "bad." The financial press of the 1950s and 1960s consistently referred to the trade surplus as the healthiest component or "one bright spot" in the balance of payments. It was therefore with some shock that the first sign of a shift toward consistent trade deficits was noted. The trade balance showed signs of "deterioration" as early as 1968–1969, when the economy overheated in response to fiscal programs, military and domestic, and accommodating monetary policies, causing an inflow of imports. Not all economists were surprised by the erosion of the trade surplus. Many thought that such a shift was inevitable. Sooner or later the United States would enter the stage of the "mature economy," in which more money flows into the economy from interest and dividends on previous foreign investment than flows out to build up new assets abroad. To make room for these inflows, deficits would have to be run on the current account. The transition was more rapid than anticipated, however, and it also had less to do with a process of natural evolution than it did with the appearance of a new player on the international scene—OPEC. The oil- and capital-rich nations of the oil cartel saved rather than spent their monopoly profits, investing them in the U.S. market. Much to the delight of the American consumer, the dominance of OPEC turned out to be only temporary, but the volatility in the balance of payments did not disappear so easily. As the financial press and yet another crop of presidential contenders have made us well aware, the 1980s saw another historic swing in the balance of payments. Even more rapidly than the transformation from debtor to creditor nation at the end of the last century, the United States had become, by 1986, the largest debtor nation in history.

Considerable handwringing has accompanied this development, but the response that "damn foreigners" be kept out of American business is not new. Indeed, the desire of producers to beat down competition by legislative restriction is as old as the market itself. Just as the mid–1980s heard a rising chorus demanding justice by tariffs, so the mid–1880s heard the call of nativist sentiment out West. When it was claimed that 21 million acres of western land had been bought up by British noblemen alone, with uncounted millions purchased by mere commoners, the Anti-Alien Act of 1887 restricted foreign ownership of land. Not every legislator climbed aboard the bandwagon; wise heads predicted that the Act's passage would reroute foreign capital to Mexico and South Africa, where it would help promote greater competition against American mining interests. That is precisely what happened, and the effect was disastrous: Capital investment in the mining industry of the Far West dried up, while municipal improvements slowed down for lack of funds. By 1900, the effects were so clearly seen that an amendment was made that permitted foreign investment in the construction of dams, reservoirs, and irrigation ditches, as well as in the all-important mining sector. The importance of foreign capital to American economic development, as well as the dangers of simplistic solutions to complex issues, is nowhere better illustrated than by this sorry legislative episode.

Deficits need not be a sign of weakness. Those of the mid-nineteenth century were a vehicle for growth and greater prosperity. Let us remember, however, that history is not always the most straightforward of guides to the present. Things have changed since the days of Andrew Carnegie and Horatio Alger. Both the gold standard and the dollar standard are dead, and the financial world now deals in floating—rather than fixed—currencies. The marketplace, not government edict, determines the value of the dollar. But once again, the patterns of the past have created a misleading beacon for the future. If anything fueled the alarm over the trade deficit in the early months of the 1986 "crisis," it was the fall of the dollar's value on the international exchanges. Unhappily for clear-headed thinking on the deficit, the dollar has become a symbol of national economic strength, when it ought to be no such thing. Currency depreciation is simply the textbook way in which trade imbalances are resolved in a floating system. There is absolutely no reason to believe that the falling dollar betrays domestic weakness (after all, when it was higher so too were unemployment and inflation), nor that it will fail to do its job.

What exactly is that job? To entirely eradicate the deficit? To ensure that the current account always balances at zero? Surely not. For one thing, the dollar could not manage it alone because the balance of payments is set by broader economic forces, at home and abroad. The current account deficit, to repeat, is equal to the gap between domestic savings and domestic investment. In the nineteenth century, the deficit resulted from an inadequate volume of savings to meet the investment needs of an industrializing nation. In the 1980s, the gap again widened, partly because the savings rate was so low—in 1986 the household savings ratio was 3.9 percent, compared with 12.7 percent in Germany, 17.0 percent in Japan, and 11.3 percent in Canada—and partly because public *dissaving* was so large. Unless and until private savings increase and/or public dissaving decreases, and as long as foreigners are prepared to lend to Americans in return for interest rates that are swollen by the internal savings–investment gap, this external deficit will endure. To the extent that foreigners are lending us money for capital projects that assist our long-term economic growth, we should welcome the situation rather than rebuff it.

Let us not be too complacent, however. Surely there are *some* dangers in the deficit? Concerned commentators in the mid–1980s were writing of "evaporating confidence" in the dollar, of a "financial crisis" leading to "a serious world recession." The fear was that a sudden and significant loss of confidence in the value of the dollar, predicated perhaps on the belief that the United States is overloaded with debt, would trigger a vast withdrawal of funds from the American economy, precipitating a new 1929. That particular fear, however, seems to be dissipatating as we get used to having a deficit. After all, currency traders continue to buy dollars, and foreigners will still find good bargains among American assets. Moreover, the American debt problem is by no means as bad as we used to think, mostly because the government figures, on which we all rely, understate the true values of American investments abroad (by using book rather than market valuation). It turns out that Americans still earn more income from our overseas holdings than foreigners earn from U.S. assets, which suggests that we are still some distance from disaster. The real danger is not foreign debt, hanging over us like some sword of Damocles. Rather, it is the threat of inappropriate protectionist policies that should disturb us the most.

CONCLUSION

Perhaps the simplest and best lesson to be drawn from history, and from the historical experience of countries other than our own, is that the trade deficit is not more than the surface manifestation of more elemental economic forces. In the nineteenth century, it was the pull of American opportunity for foreign capital that generated deficits; today, it is the combination of low savings and unbalanced budgets that is at work. The eradication of the deficit will not be achieved by tariffs, or by currency changes, but only by Americans changing the fundamentals of their economic behavior—by curbing their own, or their government's, spendthrift ways. If we are prepared to change neither, then we shall just have to continue living "in the red." If Americans, however, are serious about solving the trade deficit, then some choices need to be made—most notably, about the state of government finances. Indeed, it is hard not to resist the simple judgment that if the federal budget were put in order—as it should be, and soon—the trade figures would be dropped from the front pages of our daily papers and returned to where they belong, as statistical minutiae in the business press. That would be no bad thing at all.

12

The Great Depression:
Can It Happen Again?

JOHN WALLIS

Could the Great Depression happen again? American economic historians have been asked this question perhaps more often than any other, and never more frequently than in the months following the Crash of 1987. The traditional answer has been a strongly qualified *no*—that we learned our lessons in the 1930s and those old mistakes will not be repeated, as long as policymakers keep their wits about them; that the ability of the economy to resist depression was strengthened during the New Deal, and that these reforms have prevented runaway depressions since 1933. Are these reforms really enough to prevent another Great Depression? Have the policymakers really learned their history lesson? The events of October and November 1987 leave considerable room for doubt.

THE MARGIN OF REFORM

Three fundamental changes in economic institutions made during the New Deal are oft-cited reasons why we should not be worried about another Great Depression: reforms in the stock market—particularly limits on margin purchases of stock; the strengthening of the Federal Reserve System and the establishment of banking insurance through the Federal Deposit Insurance Corporation (FDIC); and a commitment on the part of the national government not to stand idly by and watch the economy go under, coupled with a growing sophistication about the workings of the economic system. All of these changes are certainly good things, but none of them is necessarily capable of preventing depression.

Raising the margin requirements to purchase stock has certainly reduced the volatility of the market. In 1929 stock could be bought on as low as a 10 percent margin: a $10 stock could be bought with a $1 investment and a $9 loan. The minimum margin rate is regulated today at 50 percent; the same $10 stock requires an investment of $5 and a $5 loan. Since the stock itself is the collateral for the loan, when the price of the stock falls the lender, usually a broker, calls in part of the loan. If the $10 stock drops by $1, the investor with a 10 percent margin must come up with another dollar—a loss of 100 percent on the investment. The investor with a 50 percent margin must come up with the same dollar, but loses only 20 percent. If a margin call is not met, the broker sells the stock to cover the investor's loan. In severe market downturns, declining stock prices snowball as margin calls are made and investors sell their stock to meet their calls.

Today's market is better insulated from market panics induced by margin selling, but that is not good news. The ultimate decline in stock prices must, in some rough way, be proportional to the original and "real" reasons why stock prices fell. An initial decline in the value of stock, say by 5 percent, caused by a public perception that future corporate profits will be lower, will be magnified into a larger than 5 percent decline if there is heavy margin trading. It stands to reason, however, that the decline will be larger if most margins are 10 percent than if most margins are 50 percent. In the Great October Crash of 1929 the market lost roughly a third of its value, just as the market lost roughly a third of its value in the Great October Crash of 1987.

If the market of 1987 really was more stable than the market of 1929 then, at the very least, the real reasons for the market decline in 1987 must have been as great or greater than the real reasons for the decline of 1929. In this sense, the economy may be in more trouble than anyone appreciates. The alternative interpretation, however, is equally disturbing: If institutional reforms in the stock market have not stabilized the market, then a New Deal improvement in the economic structure has had no effect. Either way, the stock market crash was not good news.

WHEN YOU CANNOT BANK ON IT

It is possible to put too much weight on stock market crashes as a cause of economic depression. Although the stock market crash of 1929 is popularly believed to have caused the Great Depression of the 1930s, the most likely cause was in fact the collapse of the nation's banks from October 1930 to March 1933, which caused Roosevelt's first act as president: closure of all of the nation's banks. Indeed, all the economic depressions in our nation's history may be attributed, in varying degrees, to the failure of financial institutions.

To understand how bank panics cause depression, we need to understand the process by which banks create money. It is widely known that banks create money, but how they do it seems to be known to only a few. Actually, it is quite simple. Suppose you take a $10 gold piece from under your bed and deposit it in your checking account. Both before and after the deposit you had $10 worth of money, first as gold and then as a checking account balance. The bank now has $10 in its vaults, some of which it will loan at interest in order to produce its profits. Say it makes a $5 loan to your neighbor in the form of a check, which your neighbor redeposits in his checking account. You now still have $10 worth of money and now your neighbor has $5 worth of money. The money supply has risen from $10 to $15; $5 has been created by the banking system by making a loan.

So far so good, but what happens if you both decide to go down to the bank on the same day and withdraw your checking account balances? The bank has only a $10 gold piece and will be unable to provide you both with cash. This is what happens in a banking panic when many people decide, for whatever reason, that a bank is in trouble and descend in throngs to withdraw their money.

Banks, even financially sound banks, must close their doors, temporarily freezing the assets of their depositors. More important, banks try to convert their fixed assets—the loans they have made—into cash to meet depositors' demands, driving down the value of those assets in the open market. This makes the position of all banks weaker and leads to more bank runs. Finally, banks are reluctant to make new loans, further slowing the pace of economic activity as firms find it harder to borrow and consumers find it harder to finance new purchases. Banking panics thus lead directly to lower consumption and investment, which in turn lead to higher unemployment. Those depositors who lose their jobs draw down their savings, which may lead to more bank panics. The economy finds itself in the grip of a depressionary cycle.

This cycle was broken in 1933 by federal intervention in the banking system, shutting down all banks, reopening only those banks that were financially sound, and—most important—insuring the deposits of the reopened banks through the FDIC. People who believed that their deposits were safe and insured did not rush down to the bank to withdraw their money when rumors began to fly. In one single act, the creation of the FDIC eliminated banking panics from the American economy.

Alas, the FDIC cannot eliminate bank *failures,* as we have seen so often in the last ten years, nor has deposit insurance spread beyond commercial banks and savings and loan associations. Banks still fail when the value of their assets, the loans they make, is less than the value of their liabilities, their customer deposits. Widespread bank failures are extremely unlikely without bank panics, but panics are prevented by insurance—*not* by the soundness of banks. This has an important implication for nonbank financial intermediaries, like money market funds and credit card companies.

Suppose that a money market fund run by a major brokerage house was unable to deliver cash to all of the depositors who wanted it on a particular day. The money market fund is not insolvent, but it takes time to convert even highly negotiable assets like treasury bills into cash. A rumor spreads that the money market is not honoring its commitments or, more likely, that depositors may not get their money out on the day they want it. As the rumor spreads, nervous investors start withdrawing funds, and pressure builds on the money market fund to convert its assets into cash. Another money fund may come under suspicion and the process moves more quickly. Ultimately, every depositor will be paid, but in the process the money fund will have to sell all of its assets, primarily treasury bonds and high-quality commercial loans, in the open market. This will drive down asset prices, drive up interest rates, and reduce the supply of loanable funds in the marketplace.

THE FAILURE OF THE FED

All of this may sound far-fetched, but it is exactly what happened to banks between 1929 and 1933—though, of course, it would have seemed equally far-fetched to investors in 1929. Investors then and now had faith that our central bank, the Federal Reserve System, known as the FED, would step in to prevent this kind of panic and decline in asset values. The FED does this simply by creating money and buying up the assets that the banks and money markets are trying to sell.

The FED creates money in the physical form of paper money and as bookkeeping entries in the accounts that member banks have at the FED. When the FED buys debt, by purchasing interest-bearing bonds and IOUs from the private member banks that make up the system, it issues more money. The FED can prevent banking panics by being willing to step in and buy the assets of banks that were facing runs. By providing banks with cash in exchange for their loans, the FED enables the banks to meet depositor's demands for withdrawals, stops the panic, and prevents the decline in the market value of the bank's assets.

One of the tragedies of the Great Depression is that the FED did not do this between 1929 and 1933. The FED allowed banks to fail and panic to spread without intervening with ready cash to buy up the loans of member banks. Had the FED done so, the Great Depression would almost undoubtedly have been less severe and, in all likelihood, not as long. One of the most important lessons of the 1930s is that the FED must intervene in financial markets during crises, as the FED did with alacrity in October.

A more important lesson, however, is to be learned from the reasons *why* the FED did not step into the market between 1929 and 1933. Any modern policy based on the same principles as the 1929 policy will simply reproduce the 1929 result.

There were three reasons why the FED failed to intervene. First, it believed many banks in trouble were financially unsound institutions that had overextended themselves during the stock market boom. Second, the FED was concerned about the value of the dollar on international exchange markets. To prevent a fall in the dollar it kept interest rates high in the United States relative to other countries by selling bonds and reducing the amount of money in circulation, thus aggravating the financial situation of banks that wanted to sell bonds to the FED and get cash. Finally, the FED was concerned that inflation would get out of hand, so it allowed the money supply to shrink and actually caused a deflation of 25 percent.

History has concluded that none of the FED's fears were justified, but no one has disputed that the FED's policies accomplished what the FED wanted to accomplish. Stop and think about that for a minute. Between 1929 and 1933 the FED followed the appropriate policies to keep the value of the dollar stable and to prevent inflation. The FED's policies were the right policies to pursue those goals, but those were the wrong goals. Even today, a monetary policy designed to keep the value of the dollar stable and to prevent inflation would look like the FED's policy between 1929 and 1933. Indeed, we may get the same results we got in 1933.

The problem in depressions is not inflation, but *deflation*. With a few exceptions, the price level has increased annually in the United States since 1933. But prices rose and fell throughout the 1800s, ultimately cancelling each other out: the price level in 1900 was roughly what it had been in 1800. Deflation has always been a real possibility in depressions because banks stop creating money when they stop lending. As the money supply falls, money becomes more valuable relative to other goods and services. A dollar now buys more of other goods and we have deflation. In its way, deflation is as bad for the economy as inflation.

Where inflation falls heavily on lenders, however, deflation falls heavily on borrowers who are forced to repay their loans in dollars that are worth more than the ones they borrowed. In 1929 most households had relatively small amounts of debt; in 1987 most households had large amounts of debt—from mortgages to credit cards, student loans to credit lines. A deflation of 25 percent in our modern economy would result in a much larger reduction in consumer expenditures than it did between 1929 and 1933, and correspondingly larger increases in unemployment.

The goals of the FED are critically important in preventing bank and financial failures and deflation. It is the *goals* of the FED, not merely its presence, that is important.

THE SOPHISTICATES ARE WRONG

What is true of monetary policy is also true of macroeconomic policy in general. We have much more experience and sophistication than we did in the 1930s. We now know something about how certain events like increases in the price of oil or a new foreign competitor can affect output and employment. We have a range of theories that explain how various parts of the economy work. We also have a wider range of policy tools to stimulate the economy. What has not changed since 1929 is that all policies, theories, and tools must be directed toward an understood goal if the policies are to achieve what we desire.

For the moment, let us assume the worst case: Output levels are headed down, unemployment is headed up, and the policies that we adopt in the next few years will determine whether we have merely a recession or a full-blown depression. All of the "sophisticated" views of the economy, even the modern ones, comprehend a simple fact: Depressions are caused by a reduction in the amount of things that individuals and businesses buy. This is indisputable. Another indisputable observation is that the amount of things that people are willing to buy, either as individuals or as business leaders, is based on their confidence in the future. Put these two indisputable notions together and one way out of a depression is obvious: increase confidence.

With few exceptions, pundits from the left, right, and center agree that recent crises of confidence in the stock market are the result of the federal budget and trade deficits. If confidence was shaken by high deficits, then the logic seems inescapable: Confidence should be restored by lower deficits. These sophisticates, however, are wrong. The worst mistake made by the Hoover administration was its focus on confidence rather than on economic *realities*.

President Hoover's position in 1929 was that nothing was fundamentally wrong with the economy, that the market crash was a temporary phenomenon, and that all we needed to do was wait and things would improve. As times got worse, Hoover stuck to his guns, declaring that the most important thing that the government could do was to ensure confidence in the markets by keeping its own financial house in order. Incomes fell as the depression spread; with incomes falling, national tax revenues declined and the federal deficit began to pile up.

Hoover was caught in a trap of his own making. He had declared that the recession was due to a crisis in confidence and that the government could best raise confidence by keeping a balanced budget. The growing depression, however, was forcing the budget out of balance, by larger amounts each financial quarter. Citizens who had taken Hoover at his word were faced with a situation of growing concern. He had told them to place their confidence in the government and now the government was running in the red. Those who did not believe Hoover to begin with could hardly take consolation in a president following the wrong policy at a time of national economic emergency.

The trap, once sprung, grew tighter. Hoover was willing to spend on emergency public works to stimulate demand and increase employment, but his commitment to the ideal of a balanced budget led him—like George Bush, sixty years later—to cut permanent spending programs and to push for tax increases. In 1932, at the very depths of the depression, Hoover signed into law the largest single tax increase in history.

LESSONS UNLEARNED

What have we learned from Hoover? Historians and economists alike universally excoriate him for his wrong-headed policies, but today's newspapers, news broadcasts, and budget summits are filled with calls for a reduction in the budget deficit and a reduction in the trade deficit. What will happen if we *do* reduce the deficits?

If we really are headed toward another depression, then cutting the budget deficit—either by tax increases or spending decreases—will cause the economy to decline even more rapidly. Tax increases reduce the amount of disposable income consumers and firms have to buy products, while spending cuts reduce the amount of products that the government purchases. In the long run, these policies are needed, but the short run effects are to decrease output and increase unemployment—the main reason the Reagan and Bush administrations put them off for so long.

If output in the economy were to decline, and if the rate of decline were to increase because of reductions in the deficit, then we would face the same dilemma that Hoover faced. As income falls, so do tax revenues. As unemployment increases, so do automatic increases in federal expenditures for unemployment insurance and welfare programs. The budget deficit will get worse. An administration that has schooled the public to believe, as we are being taught now, that the budget deficit is the cause of the downturn, will need only to look at the budget to see the signs of an even deeper recession ahead. Confidence will continue to fall.

The same thing will happen if attention is focused on the trade deficit. When the value of the dollar falls on world markets, the only immediate response that the administration has is to tighten financial markets, raise the interest rate, and encourage foreign investors to buy dollars to deposit in American markets. That, however, is precisely the monetary policy that the FED should not follow—the policy that it did follow between 1929 and 1933.

If trade deficits persist and Congress moves to protect American industry and jobs by restrictions on imports, other nations will respond in kind. This was the policy followed by Congress in 1930 with the Smoot–Hawley tariff. Today's economy, however, is more vulnerable to a reduction in international trade than was the economy of 1929 since so much more of our business is international. Again, if we are taught that our confidence should depend on the size of the trade deficit, then we may be left with policies that actually harm the economy in an attempt to promote confidence.

Have our country's leaders not learned enough from the experience of the Great Depression not to repeat these mistakes? Probably not. Confidence is not an economic reality, but a political chimera. It exists only in people's minds. Its fortunes, however, can be endlessly debated in the media. Since confidence cannot be measured, any policy can be construed to increase confidence if it is presented in the right light. Policies designed to promote confidence need not fail, but they can only succeed by accident.

President Reagan held dear the memory of Franklin Delano Roosevelt as a man who provided bold leadership in a time of national crisis. What Reagan forgot was that every time FDR took explicit measures to restore the public's confidence, he did it by cutting spending or raising taxes, and for precisely the same reasons as Hoover. FDR succeeded in stimulating the economy and promoting recovery only when he gave up on promoting confidence and admitted that the economy was in trouble and began doing something to help people directly.

CONCLUSION

The most important lesson to be learned from the 1930s is that in a crisis it is politics, not economics, that determines what the goals of government policy will be. Politically expedient policy can yield disastrous economic results.

Yes, the Great Depression can happen again. The safeguards that were put in place to prevent another depression are safeguards against the dangers of 1929. One safeguard, stock market reforms, has already proved to be ineffective. Our modern economy has developed many new ways that are not covered by other safeguards. Yet the media and business communities are both urging precisely the kind of Hooverian policies that aggravated the economic decline between 1929 and 1933. The public may be excused for being confused, since they are now instructed to watch the size of the budget and trade deficits as indicators of where the economy is headed.

One person who does seem to have learned the lessons of the Great Depression is Alan Greenspan, chairman of the Federal Reserve Board. In October 1987, Greenspan quickly stepped in and increased the money supply, providing order and liquidity to what might have become a very disorderly market. Let us all hope that this one lesson will be enough to keep us from another Great Depression.

13

The Income Tax:
An Idea Whose Time
Has Gone and Come?

BENJAMIN BAACK
and EDWARD RAY

Few aspects of public policy during the 1980s generated more discussion and debate than did reform of the federal income tax. At issue was the economic role of the federal government or public sector, which had been rapidly growing since the Great Depression. Now was the time, according to a new administration, to bring a halt to this historical trend. A major thrust of this effort would be to reduce income taxes. After all, no single element in the accumulation of economic power by the federal government was more important than the income tax, the means by which the increasing role of the government had been financed. Fundamental economic and political forces were now pushing for reform, forces in many ways similar to those that led to the adoption of the income tax itself in 1913. Indeed, by examining the origins of the income tax system, much can be learned about those economic forces and special interests that drive and shape the debate over how to finance our federal government.

A STARK ALTERNATIVE

It was during the War of 1812 with England that a government official first recommended an income tax for the United States. Secretary A. J. Dallas in a report on government finances proposed such a tax to generate revenue for the war effort. A similar tax had worked well for England in financing its war with Napoleon. Before Congress was able to vote upon the secretary's proposal, the war with England ended.

A half-century later, with the onset of the Civil War, Congress once again considered—and, for the first time, adopted—an income tax for the United States. Over the course of the war, two important lessons were learned about income tax. First, it was a very impressive generator of revenue. Although it was a new tax, by the later years of the war the income tax was accounting for nearly one-third of the all the internal revenue collected. Second, the incidence of the tax was very uneven across the States. Nearly 60 percent of the total income tax revenue collected came from New York, Massachusetts, and Pennsylvania.

With the war over, the income tax was ended and the political pressure for an income tax declined, but organized labor and agrarian movements continued to back the idea, and during the last quarter of the nineteenth century a coalition of special interests gradually coalesced in support of the tax as an alternative to import tariffs. During the Civil War, import duties had dramatically increased, to a level that averaged nearly 50 percent; however, they were not lowered after the war. By 1882, in response to growing public pressure, Congress authorized a commission to hold hearings on tariff reform; the commission recommended a 10–25 percent reduction in the tariff rates. Congress rejected the idea, but in 1887 President Grover Cleveland pushed tariff reduction to the forefront of the national agenda when he took the unprecedented step of devoting his entire State of the Union address to an attack on high tariffs.

While the debate intensified, other factors were at work to undermine support for high tariffs. Since the country was now becoming the world's industrial giant, it was producing more and importing fewer manufactured goods. It was also importing more raw materials; since raw materials were subject to lower tariffs than finished goods, the share of federal revenue accounted for by tariffs declined. In addition, the rapid expansion of U.S. exports and foreign investment created new special-interest groups who favored reduced tariffs and freer trade.

Meanwhile, the federal government faced a shortfall of revenue from other sources. The Populist Party, as well as some Democrats, generated considerable support by attacking excise taxes as unfair to farmers and ordinary working folk. There was a growing consensus that excise taxes should be lowered, if not eliminated; they could certainly not be raised to make up for the tariffs. Another major source of revenue, federal land sales, peaked in 1888 and thereafter declined as less land became available for sale and a land-conservation movement emerged, resulting in large tracts of land permanently set aside as public parks.

At the same time as it was experiencing problems with its traditional revenue sources, federal military spending was expanding. After the Civil War the navy had been reduced to a few wooden sailing ships and the Grand Army of the Republic had declined to a force sufficient only to police the Indian frontier. As the United States, however, became a major exporter of manufactured goods, and a player in global markets with far-flung interests in Central and South America, Asia, and the islands of the Pacific, Congress increased naval and military expenditures to protect U.S.

shipping and to keep markets open overseas. In 1880, Rutherford Hayes became the first president since the Civil War to argue that a larger navy was necessary to protect the nation's growing commerce. Over the next decade the United States undertook an unprecedented peacetime buildup of its military. Congress approved construction of a new steam-and-steel navy and established the Navy War College, as well as the Army–Navy Gun Foundry to help the navy catch up with European naval technology. The army budget tripled between 1880 and 1905, while the navy's share alone increased from 6 percent to over 20 percent of the federal budget. By the time President Theodore Roosevelt sent the Great White Fleet on its global voyage in 1907, the United States had the second-largest battle-ship fleet in the world.

As the United States engaged in a major military buildup, further fiscal pressures mounted as Congress dramatically liberalized the veterans' pension program, creating, in effect, the first federal social program aimed at income redistribution. Pension payments tripled and then doubled during the 1880s; together, veterans' pension payments and military expenditures increased from approximately one-third of the total budget in 1879 to over two-thirds by 1910.

With the federal government undertaking a rapid expansion of the military and veterans' pension programs at the same time that problems were emergent with its traditional sources of revenue, the result was the deepening federal crisis of the 1890s. Threatened by a mounting fiscal crisis, the government had a stark alternative: It could either institute a new tax, or it could reduce spending. Lacking the will to control spending, and with less income from tariffs, excise taxes, and land sales, Congress needed a new source of revenue.

GEOGRAPHIC BIAS AND THE CRITICAL STEP

With the confluence of economic forces producing an impending fiscal crisis, political pressure built for the reintroduction of the income tax that had been used so successfully during the Civil War. In 1894, the first year of a federal deficit since the war, Congress did, in fact, pass an income tax. The measure was supported primarily by the lower-income/agricultural South Atlantic; most of the opposition came from the high-income and industrial Northeast. The lesson of the Civil War income tax, for which Massachusetts, New York, and Pennsylvania provided 60 percent of the revenue collected, had not been forgotten.

One year later, the Supreme Court ruled that the 1894 income tax was a direct tax and therefore unconstitutional, but the pattern of federal expenditures broadened the protax coalition of special interests over the next fifteen years. A disproportionate share of the new spending initiatives between 1894 and 1908 were made in those states that traditionally opposed the income tax like Massachusetts, New York, New Jersey, and Pennsylvania. This was particularly the case for naval contracts, army arsenals and posts, and state militias, as well as veterans' pension payments. Once the economic interests of the industrial Northeast agreed to expand the revenue base to finance the increasing federal expenditures in their states, the success of the income tax was assured. Whether coincidental or by design, the geographic bias in military and pension expenditures had its effect. In 1909 Congress passed the Sixteenth Amendment to the Constitution in a near-unanimous vote; the amendment was adopted by the states in 1913.

The ratification of the income tax amendment was the critical step in the creation of an institutional framework that would accommodate rapid growth of the federal government; once the amendment was in place, income tax rates could easily be adjusted by Congress to pay for new government programs. Although in itself the income tax did not cause the government to grow rapidly, the process of coalition-building among special-interest groups, which had brought about its adoption, became an important mechanism for expanding the role of government. Indeed, the veterans' pension program of the late nineteenth century was the prototype for the social security program enacted during the 1930s, while the peacetime military buildup of the 1890s was a precursor to the post–World War II military–industrial complex. With the onset of the World War II, federal spending as a share of the GNP increased sharply and has persisted throughout the postwar era, during which the income tax has accounted for about 60 percent of the federal budget.

SOME STRIKING PARALLELS

Why did reform of the income tax system become a national issue during the 1980s? The parallels with the 1890s are striking. Just as in the 1890s, the composition of the federal budget during the 1980s changed significantly, with growing shares going to the military and entitlement programs. As Congress and the administration undertook the largest peacetime buildup in the nation's history, annual military expenditures rose by nearly 90 percent; as a percentage of the GNP, national defense outlays went from less than 5 percent during the late 1970s to over 6 percent. At the same time, the government increased its spending on entitlement programs. Despite the starve-the-poor reputation Reaganomics was given by many opinion elites, annual expenditures on social security and medicare more than doubled under the Reagan administration. The veteran's program— a key factor in the original adoption of the income tax, as we have seen— grew to the point where nearly one-third of the total U.S. population was eligible for benefits; the political clout of veterans was evident when President Reagan recommended upgrading the Veterans Administration to a cabinet-level department. In fact, the share of the federal budget during the course of the Reagan administration that was going to the combined military, social security, and medicare programs was virtually the same as the share of the federal budget going to the combined military and veterans' pension programs during the 1890s.

As in the case of the 1890s, growing military and entitlement expenditures during the 1980s focused national attention on federal taxes in the face of a deficit crisis. Some asserted that the stimulus to economic growth from income tax cuts would be so great that the income tax system would produce larger revenues with lower tax rates. Others took the position that while tax revenues might decrease, this would put pressure on Congress to reduce expenditures. Altogether, these arguments offered something for most everyone concerned about taxes. The result was a series of legislative measures to reduce individual tax rates, eliminate tax-bracket creep due to inflation, and expand the tax base.

Despite major reforms, however, revenue generated from income tax, as a share of the GNP, remained about the same as it did in the late 1970s, and the result was a fiscal crisis, epitomized by federal deficits. Just as during the 1890s certain economic and political forces helped generate the first federal deficits since the Civil War, similar forces working in the 1980s resulted in annual deficits rising from just over $70 billion to a peak of well over $200 billion. The deficits of the 1980s added over $1.5 trillion to the federal debt, and interest payments on the debt became one of the fastest growing federal expenditures, reaching almost half the size of the defense budget.

CONCLUSION

An analysis of federal finance during the nineteenth and twentieth centuries suggests that the actions of special interest groups have played a significant role in expanding federal spending relative to the GNP. Deficits are the result of the successful efforts of these groups to capture resources without paying higher taxes. The lesson of the 1890s is that deficits equal deferred taxes; the political resolution of the deficits of the 1890s was the adoption of the income tax. What form will these deferred taxes take to fund the deficits of the 1980s and special-interest programs of the 1990s? In the short run, possibilities include oil-import fees, a consumption tax, a value-added or sales tax, an increased gasoline tax, and elimination of the windfall oil-profits tax. What form, then, would a long-term resolution of the deficit problem take? Recent deficits have kept any major new social programs from being undertaken. There has also been growing pressure to reduce defense spending; during the second term of the Reagan administration, the rate of increase in annual military spending was cut approximately in half. We are only just beginning to understand the possible long-term implication of the deficits of the 1980s.

14

Are Government Giveaways Really Free?

TERRY L. ANDERSON
and PETER J. HILL

Imagine what would happen if the U.S. Treasury announced that on some future date it would give $100 bills to the first 1,000 people to enter the Treasury building. It is doubtful that the 1,000 people who received the money would be individuals who *happened* to wander into the building that day only to be completely surprised by their windfall. More likely, people would line up outside the Treasury to get the $100 bills. If they each believed a few knew of the bonanza, they might try to show up just a few minutes before the doors opened. If there was general knowledge of the free money, however, competition from other potential entrants would likely force people toward earlier arrival times. In the limit, each individual would be willing to forgo $100 worth of time and comfort to secure the prize.

Just as profits attract entry into markets, they would encourage people to queue up until the profits are dissipated. The $100,000 would be "paid for" by efforts to be one of the first 1,000 people in line. If the Treasury Department had simply given away the $100 bills to the first 1,000 building entrants with no prior announcement, there would have been a net transfer of wealth. Government giveaways, however, are unlikely to be random. Politicians generally gain support by targeting special interest groups, thus encouraging the groups to compete for the largess. Although the first 1,000 recipients bid for their $100 bills by waiting and thus dissipate most of the gain, the cost to the Treasury is still $100,000. The program is costly to government, yet offers little gain to those who were supposed to receive the benefits. Finally, when the competition causes an outlay by the recipients that does not produce anything but the transfer, the game becomes a negative sum. Resources *expended* in the pie-slicing game reduce productive capacity. This hypothetical example, as outrageous as it may seem, has a very real parallel in American history.

SQUATTERS' RIGHTS AND STARVING TIMES

At the time of the American Revolution, most of the land within the thirteen original colonies was privately owned. Beyond the boundaries of those colonies, however, was a vast domain, and by 1776 settlers were already pushing into these frontier lands. Who owned these unclaimed, unused tracts? How would title be established? Should the land be sold to generate revenue for retirement of the war debt, or should it be given away to settlers willing to populate the frontier?

These conflicting objectives often dominated the nation's early political climate, with Alexander Hamilton as the major exponent of land sales for revenue purposes, and Thomas Jefferson as the leading advocate of a free land policy. Hamilton's position dominated until the national debt was retired in 1835, after which the idea of giving the land to deserving individuals gradually gained favor. Even during the era of Hamiltonian land sales, however, preemption allowed many settlers to push into areas before they were opened for auction. There was tacit recognition by the federal government that "bidding" via premature entry was legitimate. Much as the giveaway of $100 bills in our hypothetical example would encourage queuing, so, too, did the government's land policy encourage squatting. Eventually, preemption explicitly allowed early settlers first right to purchase the land on which they had settled. Of course, the fact that *some* cash outlay was required meant that the policies did not make land completely free, but the effect was to transfer wealth in the form of cheap land.

As government continued to legitimize squatting, people recognized that in order to get the land they had to be the first to settle on it. By the time settlers had taken the actions necessary to secure the land, however, they were likely to receive little or no net benefit. There were substantial costs in settling new acreage: The land had to be cleared and planted, boundaries established, and a dwelling place built. Perhaps most important, there would be lost income from the occupation or land left behind. On the other side of the equation, settlers could expect a stream of future benefits from the land. If population continued to grow and transportation networks pushed westward, the land should eventually produce profitable crops and livestock. Settlers had to weigh these expected but uncertain costs and benefits, as well as the odds of being early enough to establish an ownership claim.

In terms of the country's overall productive capacity, it would have been more efficient if settlement had been delayed until frontier infrastructures had been developed. By postponing production from the land until it was more likely to be profitable, the net present value of the land's productive capacity might have been maximized. As long as competition for ownership claims required being there first, however, settlers were willing to rush to the land. Just as individuals would line up outside the Treasury Department waiting to receive $100 bills, so, too, did settlers move to the frontier before farming was profitable—simply because they knew that by waiting they would lose the opportunity to get land at a "bargain" price. Rather than maximizing the net present value of the land, premature entry and the accompanying "starving times" dissipated its value.

A GREAT GULF

With the enactment of the Homestead Act in 1862, land giveaway became an explicit U.S. policy objective. Efforts to give away the public domain were expanded and modified by subsequent laws including the Timber Culture Act in 1873, the Timber and Stone Act in 1878, the Desert Land Act in 1877, the Enlarged Homestead Act in 1909, and the Stock-Raising Homestead Act in 1916. All told, 285 million acres were transferred under the various homestead laws. Each of these laws attempted to reduce the burden on homesteaders, and to eliminate fraud that was putting free land into the hands of "speculators." In each case, however, the results were the same: People competed for the largess until profits were normalized.

Although the homesteaders did not pay a high cash price, they did have to bid in other ways; In other words, the land was far from free. The various acts required residence on the land for a specified period of time, the plowing of a certain portion of homestead, construction of irrigation systems, and the planting of trees. Economists Libecap and Johnson have estimated that "expenditures attributable to Federal restrictions and which involved real resources: agent payment, development costs, and miscellaneous expenditures" amounted to between 60 percent of the total value of lands disbursed under the Timber and Stone Act, and 80 percent of the value of lands given away under the Preemption Act. Where government requirements did not decisively dissipate the value of the land giveaway, homesteaders would, in effect, bid by premature settlement. As long as there was freedom of entry, the race to get to the land first would assure that bids were close to the actual value of the land, whether those bids were in the form of early arrival or farming activities that were not otherwise profitable.

In view of the necessity of bidding by settling before farming was profitable, it is not surprising that many of the accounts of early days on the homesteads report a period of "starving times." Because of premature entry into the frontier, two-thirds of all homestead claimants before 1890 were unable to meet the Homestead Act requirements and therefore failed to obtain title to the land. As historian Gilbert Fite has noted, there was a "great gulf between the hope of farmers, often built on blind faith, and the realities of settlement, which resulted in hardships, sacrifice, and sometimes complete failure." These difficult periods have generally been attributed to the lack of reliable information on the part of the settlers, who were moving into areas that required a different mode of agriculture than they left behind. Even if the settlers had better information about frontier conditions, however, competition for claims to future land productivity would have forced homesteaders to go through a starving time with negative returns.

Again, as in the case of the hypothetical Treasury giveaway, the fact that the settlers were receiving few net benefits from the free land did not prevent the transfer of valuable resources to some individuals. The land made available by the federal government was a productive resource, and was seen as such by both settlers and the government. Unfortunately, the actions required to receive it meant that the program was costly to the government in terms of foregone revenue and administrative overhead, and costly to homesteaders in terms of money, foregone income, and hardships. The homesteaders had to pay, even if the government did not receive the payment.

LATTER-DAY HOMESTEADING

Despite the failure of land giveaway programs to accomplish a net transfer of wealth, government policies continue to attempt giving away the proverbial "free lunch." Of course, if people are unable to adjust their behavior as homesteaders did by entering the frontier prematurely, it is possible for government to give people wealth. There are few examples, however, where there is no responsiveness on the part of transfer recipients. The ability of individuals to respond to such transfers is ubiquitous, and goes far beyond our ordinary conceptions.

An excellent modern example of homesteading for federal largess is recent American farm policy. Between 1979 and 1984, federal outlays on farm programs increased by nearly 600 percent; by 1985, over $9 billion was transferred to farmers, at a cost of $19 billion. The gain per farmer was $4,000 and the cost per nonfarm laborer was $190. These figures, however, mask the real net gain to farmers, who devised ingenious methods of competing for the transfers. Since many of the programs were tied to particular farm acres, the value of those acres rose in accordance with the benefits of the program. If such transfers came about without any lobbying effort by farmers, then the transfers would go to the original landowner in the form of higher land values, and anyone entering agriculture would have to pay higher prices for land. Farm groups spent very large amounts "homesteading" for farm programs, however, so the transfers were not free.

Once programs were in place, there were other responses to programs designed to limit production and increase input prices. Farmers took the least fertile land out of service, and added large amounts of fertilizer and other inputs to the remaining acreage in order to produce more crops with artificially high prices supported by government. In consequence, the farm program became enormously expensive to taxpayers, but the net transfer to farmers was far less than the cost to the national Treasury.

Even welfare and unemployment policies, specifically targeted at the poor, tend to have the same effect. From 1965 to 1983, total transfer payments in 1983 dollars increased from $148.6 billion to $475.6 billion. Despite this more than 200 percent increase, little progress was made in reducing the poverty rate; in fact, for those of working age, the poverty rate actually rose. For those over sixty-five, however, the poverty rate fell from 22.4 percent in 1965 to 4.3 percent in 1984. Why such different results from government transfer programs? A plausible explanation hinges on the ability of those involved to respond to requirements for receiving transfers. Since eligibility for aid was largely dependent on age, the elderly had far fewer options to change their behavior. By contrast, attempts to give benefits to those of working age on the basis of means tests for income met with much less success. People could and did "queue up" for income transfers by remaining unemployed for longer periods, reducing their earnings, and changing their marital status.

CONCLUSION

Political rhetoric aside, the state is not a very efficient engine for the carrying out of transfer policies; no matter how worthy the cause, it is difficult for government to bestow net benefits on people by the transfer of resources to "deserving" parties. Nineteenth-century U.S. land policy attempted to give away substantial amounts of "free" land, but unfortunately for those who received the land, the gift was not free at all—and probably represented a net drain on the productive capacity of the country. The problem is that government cannot give away largess, because, surprisingly enough, people insist on paying for it.

15

Do Americans Want
Big Government?

JONATHAN HUGHES

Ronald Reagan, elected president in 1981 with a mandate to "get government off the backs of the people," left behind a governmental establishment
in most ways larger than it was when he came to power. Federal civilian
employment rose from 2,772,000 to 2,988,000 in 1981–1989; federal outlays
rose from $591 billion to $1,143 billion in the same period. By some measures these increases actually measured a slowing down in the rate of growth
of federal government and that was, in the circumstances, a kind of success
for Reagan's policies. Powerful interest groups had wanted even larger
increases in governmental size. Reagan was up against a deep and puzzling
phenomenon that no amount of "communicating" alone could reverse:
Despite the disclaimers of our statesmen, the size of government has grown
because the American people demand ever-increasing governmental
functions.

ARE WE 45 PERCENT SOCIALIST?

The relative size of the federal government's bureaucracy, and its rate of
growth, have varied throughout our history. The first surge of growth, from
the late nineteenth century to the early 1920s, encompassed the years when
the modern federal government was first positioned in Washington, D.C.,
with its first agencies of nonmarket control permanently established, after
the liquidation of the World War I command economy. The New Deal
surge is well known. From World War II to 1970, including the Great
Society bureaucratic buildup, the mean annual rate of growth slowed down,
even though relative growth continued by this measure. At the peak,
around 1970, federal employment had grown some 3.5 times as rapidly as
the population had grown since the 1920s. Then, in the 1970s, as computers

replaced man/woman power per unit of work done, the number actually declined sharply. The Reagan administration managed to contain further per capita growth of the bureaucracy.

Viewed as expenditures, the federal government has grown even faster. Since 1929, government expenditures have grown four times as fast as the GNP, and the federal share alone has grown seven times as fast. In 1929, the federal share of the GNP was only 3 percent. By 1987 it was between 20 and 25 percent. Since 1950, growth of federal expenditures, inflation included, has been far more rapid than the economy's growth. From 1950 to 1980, federal expenditures grew about half again as fast as did the GNP. After 1960 the federal share surpassed the combined shares of state and local governments and has remained so to this day.

What this has meant for the nation's "agenda" for the future is striking. In 1929 private investment was five times as great as the federal budget; by 1991, it was only about half as much. Total government expenditures, federal, state, and local combined, now distribute some 45 percent of the GNP, whereas that figure had been only 12 percent in 1929. The private sector has only 55 percent of the GNP within its control. Are we 45 percent socialist?

ESCAPING THE MARKET

Another important dimension of governmental growth, the presence of regulation, is not easily quantified. Even if there were some way to count up all the regulations, and there is not, their effectiveness depends not only on the degree of compliance, but also secondary effects, backward and forward linkages from regulation-constrained activities. What we do know is that we have an economy that is *different* from the one the free market would produce.

A century ago there was little federal regulation of private economic activity; today, federal nonmarket control of the private sector is very nearly ubiquitous. Even a regulatory program so designedly circumscribed as our antitrust legislation affects some 85 percent of all measurable economic activity.

Someone wants, or once wanted, each regulatory activity in order to escape from the results of market decisions. Among the results have been rent-seeking and free-riding activities, which, inevitably, have created a clientele that opposes reform or reduction of regulation. Hence, in 1987 we celebrated the centenary of the Interstate Commerce Commission (ICC) despite the difficulty in discerning what good it does for the economy, or ever did.

Change in regulations for some financial institutions, the airlines, and interstate transportation during the Reagan years, for example were a drop in the bucket. There is no reason to suppose that the extent to which each person's life is subject to government regulation has not continued to increase in the past seven years, as it has for the past century. Moreover, following the financial debates of 1989–1991 there is a powerful outcry for *reregulation*. We do not even trust the free market at the center of American capitalism—its financial heart.

HORATIO ALGER AND THE WELFARE STATE

There is no general agreement about the causes of big government in this country. The developed countries of the world (i.e., those who consider their economies to be modern) have proportionately large-scale government sectors; many of them are larger in that respect than ours. The developed countries all have extensive welfare states, and that, to a large extent, accounts for the sizes of governments. Since no welfare state exists in an industrial country in the presence of *small government,* one cannot just test to see if welfare states necessarily make governments relatively big. Such a condition, however, seems to be logical enough.

It could be argued, of course, that welfare states exist *in order* to magnify the sizes of government sectors. Watching the bureaucratic source of resistance to *Perestroika* in the former Soviet Union underlined this idea. One does not suppose that governments are dragged, kicking and screaming, into bigness in order to bestow modern welfarism upon a grateful citizenry. Whatever good welfare does for the population, it does at least as well, by providing jobs and incomes, for the dispensers of welfare-state largesse.

The fact is, however, that government grew mightily from 1870 to 1932, before there was any significant federal welfare state in this country. In other words, the American version of big government predated the welfare state.* A key period was 1887–1890, which saw creation of the ICC and passage of the Sherman Antitrust Act. Over the following two decades the Food and Drug Administration, the Federal Trade Commission, the Federal Reserve System, the Federal Income Tax Amendment of 1913, and much else would be added. As late as 1886 none of that had existed, despite the growth of governmental activism in the state legislatures. There was as yet not a hint of a federal welfare state, except for pensions for veterans of the Civil War.

It has been argued by one observer[†] that big government has grown in this country primarily because of two related forces: the ratchet effects of

---

*James Bryce, *The American Commonwealth* (1893) carefully considered the (then) recent expansion of state-level regulatory activity. He considered it both contradictory to common American ideals and economically misguided—Why undo America's great and original creation, the late-nineteenth century laissez-faire economy? Clearly that was being done by state legislatures and Bryce was puzzled by it. He even went so far to suggest that Americans were naïve, and believed they could get rid of government as easily as they could create it. Modern European history had already proved how mistaken such ideas were.
†Robert Higgs, *Crisis and Leviathan: Critical Episodes in the Growth of the American Government* (1987).

twentieth-century crises and an adaptive ideological transmutation in which Americans abandoned the heritage of sturdy individualism and came to embrace the welfare state. Undoubtedly, the growth of big government can be traced to past crises, such as the Great Depression of the 1930s. The extent of popular commitment to rugged individualism in the past, however, is unmeasured and probably exaggerated.

Foreign visitors from at least Jacksonian times onward (Mrs. Trollope, de Tocqueville) remarked on the spirited individualism expressed by ordinary Americans. On the other hand, as Lord Bryce noted, there has also existed a naïve and perhaps contradictory willingness to utilize the coercive power of government to achieve all sorts of redistributive objectives.

Even before the rise of the Populist Party in the late 1880s there were tariffs, subsidies, special franchise monopoly grants by the hundreds, monopoly powers given out for canal and railroad companies, and federal land sales that were irregularly but effectively controlled by the Northwest Ordinances. Then, as in colonial times, lands were granted for services under the 1862 Homestead Act. Huge land grants, 10 percent of the entire public domain, were made to finance the transcontinental railroads, and extra land grants were designed to promote irrigation projects in the west. Indeed, there are many examples before 1890, at the federal level, of a readiness by special interest groups to be "collectivist" to achieve desired objects that lay beyond the grasp of unaided individual enterprise.

The problem has not been liberal or ideological change, but conservative adherence to an old custom. The *ease* with which we slipped into the modern welfare state, after all, can only be due to something fundamental in our institutional legacy and history. We developed a certain precocity in the use of institutions of democratic coercion from colonial times onward, but for long decades these actions were largely restricted to municipal law and to the state legislatures. In modern times, we lump these practices and institutions together under the rubric *police powers,* which is the unquestioned right of governments to intervene in private economic life for the protection of public health, safety, and welfare.

In colonial times exercise of these powers comprised the main substance of colony governments,* but when the U.S. Supreme Court held in 1877 (*Munn* v *Illinois*) that any private property "clothed in the public interest" was lawfully subject to government regulation, it was to the police powers that the court mainly appealed. In 1934, in *Nebbia* v *New York,* when the Supreme Court opened the floodgates to regulation of *any* private property at all, discarding the limitation of *Munn,* the police powers were again held to be sufficient justification.

Although the doctrine of rugged individualism remains an American ideal, we have grown accustomed in our daily lives to the use of government

†Jonathan Hughes, *Social Control in the Colonial Economy* (1976).

coercion to achieve special interest and, more rarely, public policy objectives. Congressmen, needing votes, have become increasingly easy prey to organized and clamoring economic coalitions who invoke the general welfare as justification for special-interest legislation.

DOOMED TO DISAPPOINTMENT

Big government was what we wanted and what we got. Yet we, or some of us, have been soured by the failure of our creation to deliver the goods, by its huge costs, by its feckless incompetence, and by its flamboyant episodes of power abuse. These things have given neoconservatism new life after four decades of welfare statism. As the Reagan administration quickly found, however, it is not within us to abolish our welfare state. We are hooked on it, and it poses a terrible dilemma.

The institutions and habits of nonmarket control have always been present at the state and local government levels, where they are often flexible, amenable, friendly, and congenial. In raising those powers to the federal level, however, we have created a monster, and we have not found a way to reform it or even to control it.

If our motives seemed always constructive, reformist, and benign, then the outcomes, too, have been often unexpected and disappointing. The ICC presided over the dissolution of the passenger railroad system it was supposed to perfect. The liberated Federal Reserve System backed the mounting federal deficits with money creation that produced inflation and balance-of-payments bedlam. The Post Office lost money, could not deliver, and has ended up being a helpless fugitive hiding behind its first-class mail monopoly. The Synfuels Corporation became a haven for money-consuming boondoggling. Federally sponsored and protected nuclear power spreads a contagion of bankruptcy through the nation's power grid, as well as perpetual radiation contamination in its waste products. The income tax amendment produced a nation of tax avoiders, and a paradise for lawyers and public accountants. Medicare and Medicaid, designed to assist the elderly, also produced alarming scandal and corruption. Quotas on foreign auto imports, designed to encourage domestic industrial efficiency yielded even more expensive American autos and huge bonuses for the industry's executives, while auto plants were closed up and workers turned into the streets. A half-century of farm policy manufactured unparalleled financial catastrophe for the nation's farm families, while most of the Commodity Credit Corporation's funds aided only the largest agricultural firms. Aid to Families with Dependent Children (AFDC) produced intergenerational welfare dependency among the nation's urban poor. Decades of legislation to force virtue upon Wall Street has been followed by ever-more sensational cases of greed illegally nourished by corruption and insider trading. Other examples are in endless supply.

CONCLUSION

Responsibility for the growth of American government must rest with us, and not in our stars. We are altruists, predominantly Judeo-Christians, and we want outcomes that the free market will not produce on its own. We turn to government to get those outcomes. What we get is something different, and it usually has been quite a shock. Judging from our history, we will never depend entirely upon free contracting to allocate wealth, income, and goods and services. We have not found a way, however, to reallocate what we desire through the governmental methods we have invoked.

V
REGULATION, DEREGULATION, AND REREGULATION

The deregulation of the late Carter and early Reagan years seems to have taken place in a galaxy long ago and far away. Today, the cry is heard on all sides for reregulation of airlines, cable TV, banks, and any other industry in which capitalism does not create heaven on earth.

Wait a minute. Give a second thought to free banking, free prices, and free securities markets, and to the essays by Richard Sylla, High Rockoff, and Susan Phillips and J. Richard Zecher. Was regulation in fact all it was "designed" to be?

16

Should We Reregulate the Banks?

RICHARD SYLLA

The financial crises of the 1970s and 1980s—the failures of banks and savings and loan (S&L) associations, the stock market crashes, the junk bond defaults, the collapses of securities firms, and the scandals associated with these events—have revived one of the oldest debates in American history. Should financial institutions, the holders and guardians of most of our money and savings, be regulated in order to assure that our assets are safely managed? Or are such regulations themselves a threat to the safe management of assets? These questions were certainly asked long ago, and on more than one occasion, but the answers given in earlier eras invariably led to reactions in the opposite direction at a later time. Hence, from first to last, our financial history has been marked by cycles of regulation and deregulation.

THE POLITICS OF BANKING

A cycle of banking regulation quite similar to that of recent decades occurred in the United States a century and a half ago. In the early decades of the republic, the business of banking was regulated, mostly by state governments. Banks then were regarded much as public utilities are regarded today. They received corporate charters by individual acts of the state legislatures; these special charters gave the banks exclusive privileges, rather like the power company and the cable TV franchises have now. In return for grants of exclusive banking privileges, state governments demanded both financial considerations from banks and the right to regulate them.

As the U.S. economy grew and developed in new directions, banking under exclusive privileges became quite profitable. These profits were a signal for new competitors to enter banking. Organizers of newly proposed banks, therefore, applied to state governments for charters. If these organizers were not well-connected and loyal to the political parties in power, however, new charters were often denied—unless, of course, above-board or under-the-table payments were made to the states or to influential state legislators. Established banks also played this game by attempting to stave off new competition with similar payments to politicians and public officials. Politicians themselves were far from passive referees in the competition for banking privileges and profits. They had their own resources and needs. Perhaps those in charge of the government finances would strengthen their political power by placing public funds in the "right" banks, where the money could also be used for private profit. Thus, President Thomas Jefferson wrote to Treasury Secretary Albert Gallatin in 1803 that "I am decidedly in favor of making all the banks Republican by sharing deposits among them in proportion to the dispositions they show." The politics of banking was carried on at the highest levels.

If would-be bankers could not obtain state charters to enter the business, there was an alternative path to profits. In the old-world tradition it was not necessary to have a charter in order to be a banker. Indeed, in Great Britain and on the continent of Europe, banking with a corporate charter granted by a government was the exception rather than the rule. The rule was private, unincorporated banking. It did not take long for early Americans to adopt this alternative. The established chartered banks, however, did not approve of this unauthorized competition any more than they approved of extensions of authorized competition. They therefore asked their state legislatures to ban or circumscribe private banking. During the first two decades of the nineteenth century many states responded to this request from the established banks by enacting restraining laws to prevent or limit unauthorized banking.

Another effect of the mixing of banks and politics in this era was the liquidation of the First and Second Banks of the United States. The First Bank, proposed by Treasury Secretary Alexander Hamilton on the model of the Bank of England, was chartered for twenty years by Congress in 1791. Owned partly by the federal government and partly by private investors, it served as the government's banker while also carrying on a commercial banking business. Because of its large size compared with other banks, its having branches in various states, and its relationship to the U.S. government's finances, the First Bank was able to influence, to a limited extent, the money-creating activities of the state-chartered banks. Although the federal bank appears to have discharged its functions quite well, the state banks resented its competition and its regulatory capabilities. They used their own influence to persuade Congress not to renew the First Bank's charter, and the federal bank expired in 1811. Without the bank, however, federal finances became chaotic during the War of 1812, and one of the first tasks of Congress in 1816, after the war, was to charter a Second Bank of the United States, again for twenty years.

THE FREE-BANK EXPERIMENT

The Second Bank was like the First, except larger and with more branches in more states. After 1823, under Nicholas Biddle, its third president, the bank became an embryonic central bank, stabilizing the monetary and banking systems while continuing to compete with the state banks for commercial business. The old resentments against a federal bank arose again. Congress voted to renew the bank's charter in 1832, but President Andrew Jackson's veto of the act was not overridden, and the federal bank expired in 1836. The second of two experiments in monetary control and nationwide banking, like the first, had run afoul of politics. Two consequences of the Second Bank's demise were that the United States would not have a central bank until 1914, and would be without interstate branch banking until the 1980s.

Meanwhile, the old system of chartering banks by special acts of state legislatures became increasingly unpopular. In the interest of existing banks, it had served to stifle competition. In the interest of politicians, it had tended to promote a certain amount of corruption and influence peddling.

By the 1830s, a new approach was put forward that espoused greater democracy in banking politics and freer competition in banking enterprise. A cogent statement of this new approach came from Treasury Secretary Roger Brooke Taney in 1834:

> There is perhaps no business which yields a profit so certain and liberal as the business of banking and exchange; and it is proper that it should be open as far as practicable to the most free competition and its advantages shared by all classes of society.

This was quite a change from the earlier view that banking was properly an exclusive privilege.

In practice this new way of thinking was institutionalized, starting in the late 1830s, in so-called free banking laws. Free banking changed the chartering of a state bank from something that required a special legislative act to a routine administrative function of the executive branch of government. Those who wanted to enter the business of banking did not have to overcome the opposition of existing banks or make payments to states, politicians, or legislators. A century later, historian Bray Hammond said of these laws that "the result was that it might be found somewhat harder to become a banker than a brick-layer, but not much." By the 1850s, state after state had adopted a free banking law, and a nationwide banking system based on correspondent relationships between banks in different states began to emerge.

To protect holders of bank liabilities from losses due to bad management, a key feature of free banking was the backing of banknotes by U.S. and state government bonds. In the free banking era before the Civil War the

United States did not have a uniform paper currency such as it has today; instead, banks issued their own currency notes, which were redeemable at the banks in gold or silver coin. The analogous procedure today is cashing a check. The banks held reserves of gold and silver to meet day-to-day note redemptions, but these were fractional reserves: Since paper currency was more convenient to carry around and use than the equivalent in coins, ordinary demands for redemption could be met with a reserve that was only 10–20 percent of the note issue outstanding. Suppose, however, that redemption demands became extraordinary, or the bank failed. To ensure that noteholders would not suffer losses from such events the free banks were required to purchase government bonds as backing for their note issues, and to deposit these bonds with a state authority. Then if for any reason the bank itself could not redeem its notes, the bonds would be sold and the proceeds used to compensate the note holders.

On the whole, the verdict of modern scholars who have studied the free banking experience is that it worked quite well.* Many new banks were organized and entered the banking business, and the U.S. economy grew rapidly in the free banking era from the 1830s to the 1860s. Free banking was popular, as evidenced by its adoption in state after state and by its adoption in a modified form—the National Banking System—at the federal level during the Civil War. What made free banking so attractive to the U.S. government during the Civil War was the aid given to government finance: Backing banknote issues with government bonds increased the demand for government debt, making it easier for states to borrow, which they wanted to do in order to build transportation facilities and make other internal improvements. Best of all, free banking avoided the institutionalized graft of the state-chartered regulatory alternative.

THE WILDCAT PROBLEM

Despite these beneficial effects, the free banking system had some problems. By making it relatively easier to enter the banking business, the system attracted some persons with few skills and not much experience in banking. As a result there was a fairly high rate of appearances and disappearances among free banks. This was not bad in itself, but it led to losses for bank investors and creditors when a lack of managerial skill and experience caused free banks to fail.

The most notorious problem of free banking was so-called wildcat banking. Wildcat banks were institutions organized not to engage in a full banking business, but merely to make quick profits from putting banknotes into circulation when it was profitable to do so under imperfectly drawn

*See, for example, Hugh Rockoff, "The Free Banking Era: A Reexamination," *Journal of Money, Credit, and Banking* 6 (May 1974), pp. 141–67, and Arthur Rolnick and Warren E. Weber, "New Evidence on a Free Banking Era," *American Economic Review* 73 (Dec. 1983), pp. 1080–91.

or administered free banking laws. Such banks were called *wildcats* because, not intending to redeem their notes, they located their businesses in remote locations thought to be inhabited mostly by big-clawed cats. The opportunity to profit from wildcat banking arose when the bonds used to back note issues could be purchased by the bank for less than the value of the notes to be circulated; the difference in these values was a profit to the wildcat banker. When such notes could not or would not be redeemed at the bank—assuming the bank could be located—the sale of the bonds would not result in proceeds sufficient to cover the value of the notes, and noteholders suffered losses.*

The solution to the problem of wildcat banking was fairly simple: Legal and administrative authorities had to specify that the market value of the bond collateral backing notes must be greater than or equal to nominal value of the notes circulated at all times. For example, suppose a bank bought a bond for $1,000 and used it to back $1,000 of the bank's notes. If the bond subsequently declined in market value to $900, then the bank should have been required to reduce its note issue by $100, so that the remaining $900 of banknotes would be backed by an asset—the government's bond—worth $900. Had such a measure been consistently enforced, wildcat banking would have been impossible and losses to noteholders from wildcatting or from declining bond values would have been avoided.

Although there is a lot of folklore about wildcat banking in the free banking era, its significance for the operation of the U.S. banking system and banking's role in the economy was minimal. The losses suffered by noteholders under free banking were not large in the overall scheme of things, and wildcat banking accounted for only a part of them. The deregulated free banking system before the Civil War worked quite well, which is why it was adopted by many states and, in 1863, by the federal government.

NO BANK SHOULD FAIL

The disaffection with the old system of regulated banking that led to the movement for free banking in the 1830s bears a certain resemblance to the disaffection with regulation that led to banking deregulation in the early 1980s. In this century, the old system of regulation resulted from legislation enacted in the wake of the Great Depression of 1929–1933.

---

*A decline in the value of bond collateral between the time bond-backed notes were issued and the time when the bonds had to be sold to redeem notes could have the same consequences even in the absence of an intent to engage in wildcat banking. The greatest losses to noteholders from wildcat banking occurred in the midwestern states of Michigan, Indiana, and Minnesota, but even in New York, the state with the longest free banking experience and negligible wildcatting, the losses to noteholders were nearly $400,000 between 1838 and 1860. In the United States as a whole, the losses through 1860 have been estimated at nearly $2 million, with more than half of the total occurring in Michigan in 1837 and 1838.

American banks had failed by the thousands in these years. For a variety of reasons the banks themselves received much of the blame for the disaster. The result was a heavy dose of regulation. The Banking Act of 1933 separated commercial banking, the taking of deposits and making of loans, from investment banking, the underwriting of and dealing in stocks and bonds, on the grounds that the commingling of these two activities had encouraged banks to take excessive risks. Another provision of the act, Regulation Q, prohibited payment of interest on checking deposits and gave the Federal Reserve System, the central bank established in 1914, the power to regulate interest on time deposits because that price competition for deposits had encouraged banks to take excessive risks in lending. The 1933 Act also established the Federal Deposit Insurance Corporation (FDIC) to insure bank deposits up to a maximum amount for an insurance premium that was a flat percentage of deposits held in the banks. The objective of the act was to make banking safer by making it less competitive.

For the next thirty years, until the mid-1960s, American banking was mostly a dull, safe, noncompetitive, and highly regulated business. The only innovative area was in international banking, but a relatively small number of large, money-center banks were involved; significantly, this was the one area where U.S. banking regulations held minimal sway. The functional concept of regulation in these years was that no bank should be allowed to fail—which, in practice, meant that no bank should be allowed to take much risk, compete for deposits, or be threatened by the competition of new banks.

Beginning in the later 1960s, rises in market interest rates, caused by government borrowing for Great Society programs and the Vietnam War, led to increased inflation and a periodic draining of deposits from banks. The worst effect of this was a credit crunch—that is, a temporary drying up of bank credit to many customers. These credit crunches were but a foretaste of the more chronic condition that developed in the 1970s, when inflation and market interest rates kept on rising.

By then, the money market mutual funds had entered the scene. These funds sold shares in a portfolio of short-term money market instruments that yielded a rate of return well above what the government regulators allowed the banks, the S&Ls, and other thrift institutions to offer their depositors. Because the assets of the money market funds were short-term liquid instruments, their shares did not fluctuate much in value. Moreover, the technology of money transfers had improved to the point where the fund could offer instantaneous deposits and withdrawals, and even the service of third-party payments by checks, which the funds willingly supplied to their investors. All one had to do to earn 9 or 10 or 11 percent in a money market fund and enjoy its other conveniences was to take money out of a bank checking account paying no interest or a thrift account paying a regulated maximum of 5 or 6 percent interest and place it in the fund. Such a deal was hard for Americans to resist.

Money market fund assets grew from nothing in the 1970s to $230 billion by 1982.

While this lesson in circumventing regulation was pleasant enough for money holders, it caused problems for the banks and thrifts. The money market funds were like the private bankers of the early nineteenth century: They were unregulated competition "preying on" the regulated sector. As funds were shifted from banks to money market funds, the bankers, who both lent and borrowed for the short term, were inconvenienced and lost market share. The thrifts were hit harder. They borrowed short term, in the form of deposits, to lend long term, typically in the form of home mortgages. As market interest rates soared with inflation, the thrifts lost deposits to money market funds and saw the value of their loan portfolios collapse. By 1980, many were on the verge of bankruptcy.

Congress responded to the banking problems of the 1980s just as states had done in the free banking era, with deregulatory legislation. The Depository Institutions Deregulation and Monetary Control Act of 1980 began a phase-out of Regulation Q, which was completed in 1986. It also allowed thrifts to make a wider range of loans, including adjustable rate loans, and to offer checking accounts paying interest. In essence the thrifts became banks. They were like the free banks of earlier history that entered into competition with the old chartered banks. Another federal law, the Depository Institutions Act of 1982, extended deregulation by allowing all depository institutions to offer special accounts with no interest ceilings and by further liberalizing the loan and investment options available to thrift institutions.

One key element of the regulatory system born in the 1930s was retained: deposit insurance. It protected depositors against losses if banks or thrifts failed. Indeed, deposit insurance served the same function as the government bonds that backed the note liabilities of free banks in the nineteenth century.

A SURFEIT OF RISK

What have been the effects of this most recent round of banking deregulation? Were they similar to the effects of a century and a half ago? The similarities are fascinating. In many ways greater competition in banking has created economic benefits for bank customers and for the economy, just as it did in the past. The customer earns a market-determined yield on funds placed in a bank or thrift, and a wide variety of loans are available. Credit crunches and disintermediation are not the problems they once were. After 1982, the U.S. economy grew without the interruption of recession throughout the 1980s and into the 1990s, one of the longest sustained economic expansions in its history.

Just as with free banking, however, the latest round of deregulation has created problems. Not much in the experience of S&L managers in the highly regulated environment before the 1980s prepared them for the surfeit of risk they faced when federal law changed. Many bad loans were made, and consequently many S&Ls failed.

There are easy parallels with wildcat banking. In more than a few documented cases, shrewd operators gained control of banks and S&Ls and made questionable loans and investments. These were always funded by deposits that were attracted by paying deregulated market rates and were insured by the government. Banks and S&Ls also invested in many of the questionable "junk bonds" that proliferated in the 1980s, which offered high yields along with risks that were not always appreciated. These junk bonds were analogous to the state government bonds of questionable value that wildcat banks had used to support their notes. If junk bonds fell in value, the solvency of the institutions that held them was threatened.

Yet there is one major difference between the problems of the free banking era of deregulation and those of our time. The losses under free banking were small and did not do much to undermine the popularity of the competitive banking system, whereas the losses suffered from recent bank failures—and, especially, the hundreds of billions of dollars lost in the S&L debacle—have led many to question the wisdom of deregulation itself.

A key factor between the two eras of deregulation that accounts for the difference in loss experiences, however, is the comparative incompleteness of our recent deregulation. The regulations of the 1930s established nationwide deposit insurance in U.S. banking, including S&Ls; it was the residual presence of this federal regulation that led depositors, unquestioningly, to place their funds in the banks and S&Ls that bid for them in the 1980s at deregulated market rates. By contrast, in the free banking era the system was not so centralized. State government bonds backed the notes of free banks, but the value of the bonds, and, hence, of the banknotes, was continuously monitored—not only by the state authorities but, more important, by the market in the form of redemption activities: presenting banknotes for payment at the banks, and by banknote reporters publishing information at frequent intervals that informed the public of the value of the notes of various banks.

In the recent period, federal deposit insurance seemed to make this sort of market monitoring unnecessary; the function was left in the hands of governmental authorities charged with monitoring the activities of the institutions whose deposits they insured. Responsibility for the recent debacle, which will cost U.S. taxpayers dearly in the years ahead, therefore belongs at the doorstep of these authorities rather than at the doorstep of deregulation. These authorities failed to make the cost of deposit insurance commensurate with the risks taken by the banks and thrifts. As a consequence, the losses from insolvencies and failures were far greater than were deposit insurance reserves.

CONCLUSION

The task for our time is not to decide whether regulation, deregulation, or reregulation is the road to take. Rather, it is to determine the regulatory arrangements that assure the safety of bank liabilities held by the public and the responsibility of the financial system to its clients while allowing latitude for the markets to promote efficiency, innovation, and economic growth. To do this will not be an easy task. The task, however, will be easier if we embark on it with a better understanding of our financial history. The historical experience of free banking suggests that a deregulated banking system can function well provided that the risks banks take are continually monitored—that is what presenting banknotes for redemption accomplished—and that knowledge of these risks is widely disseminated—that is what bank-note reporters did. If deposit insurance is maintained as the modern equivalent of bond-backed banknotes, then the insurance premiums will have to reflect the risks banks take. That will require continual monitoring of bank lending and investment activities by the government authorities charged with overseeing financial institutions, with attendant publicity to inform the public, as bank-note reporters furnished long ago. This is not reregulation. It is making deregulation work.

17

Can Price Controls Work?

HUGH ROCKOFF

Whenever confronted with spiraling wages and prices during periods of inflation, policymakers are often tempted to freeze wages and prices by force of law. Opponents of such government intervention are quick to cite President Richard Nixon's dismal experiment in 1971. Interventionists insist, however, that price control *did* work in World War II, so who is right? In this chapter, we will see what history has to say.

COLONIAL CONTROLS

Controls have been used off and on for centuries. In the United States there have been attempts to control prices since the earliest days of European settlement. One of the first was a proclamation setting maximum prices on a list of commodities issued by the governor of Virginia in 1623. Later, during the Revolutionary War, the several states issued lists of maximum prices when faced with hyperinflation after the Continental Congress printed an excess of paper currency. There were also scattered attempts to limit prices at the local level. Although there is room for further research on these episodes, there seems little doubt that these attempts ended in failure. Efforts that were local or regional led to a diversion of supplies to regions where prices were free. Revolutionary merchants, even in Philadelphia—the seat of the Continental Congress—were not above evading controls. Governments that did not have the resources to collect adequate taxes were not in a position to set up a bureaucracy to administer and enforce controls. With prices in Philadelphia rising by a factor of more than 200, there was simply no hope that the limited efforts made during the Revolution could contain the inflationary pressure.

WORLD WAR I AT HOME

Except for a few scattered attempts, price controls were not used during the Civil War, but in the twentieth century controls were used during each wartime inflation. After America entered World War I inflation was severe, moreso than can be accounted for simply by the monetary and fiscal expansion. It is not surprising, therefore, that a number of agencies were created to control prices. Most prominent were the Food Administration under Herbert Hoover, the Fuel Administration, and the Price Fixing Committee of the War Industries Board. The formation of these agencies occurred along with a cooling of the inflationary fever, and controls may well deserve part of the credit. The war ended a little more than a year after controls were imposed, but there were already signs of some of the long-term problems associated with price controls.

In the industrial sector, the government tried to substitute a priorities system that required a manufacturer to fill an *A* order before a *B* order. After the war Bernard Baruch, the head of the War Industries Board, hailed the priorities system as a great success, but even during its short use in World War I the system exhibited a problem that became acute when it was revived in World War II: priorities inflation. The system was flooded with high-priority contracts, and the government was forced to issue higher and higher priorities to preserve a semblance of order.

After price controls were lifted there was no price explosion; prices moved upward, but this postwar inflation probably owed more to postwar monetary policies than it did to pressures built up under wartime controls.

FATTY HAMBURGERS AND COARSE YARNS

The longest American experience with controls was during World War II. Initially, the government tried moral suasion to control prices, but this policy was unsuccessful. The next step was to rely on controls on "strategic" prices while allowing less important prices to remain free. This policy also failed, as prices in the uncontrolled sector surged ahead, responding in part to a shift in demand from the controlled sector. Eventually, across-the-board controls prohibiting most price increases were imposed under President Roosevelt's famous "Hold-the-Line" order of April 1943. Like the less radical efforts in World War I, this policy seems to have had the initial effect of calming the economy and slowing inflation.

The monetary and fiscal system of World War II, however, was set on a course destined to undermine the controls. A substantial part of the war was financed by selling bonds to the Federal Reserve, thus producing a rapid increase in the stock of money. There were classic examples of black markets, but other forms of evasion were probably even more important. Butchers added more fat to the hamburger, candy makers reduced the size of their bars and used inferior ingredients, clothing manufacturers used coarser yarn, landlords refused to make repairs, and so on. Another frequent means of evasion was the tie-in sale. Merchants required the purchase of unwanted merchandise along with the item in demand. Less obvious was the elimination of lower-priced lines of merchandise. Clothing manufacturers, for example, eliminated lower quality products and forced people to buy higher quality, more expensive items than they would in a free market. Forced uptrading, as it was known, hit hardest at the poorest members of society.

When controls were removed in 1946, there was a classic example of pent-up inflation: Prices were immediately raised. It is probably not correct to say that in the end prices reached the point they would have in any case. In the absence of controls during the war, and any change in the fiscal system, the government would have been forced into printing more money to pay ever higher prices in order to finance the war. Moreover, postponing inflation, while it may not have seemed such a bargain in 1946, might have been seen in a different light in 1941.

ONE CASE THAT WORKED

The rapid rise in prices after the start of the war in June 1950 provides about as pure an example of a velocity-driven inflation as one can find, since money per unit of real output was almost stable during the war. Although the imposition of a price freeze in January 1951 probably should be given some of the credit for a calming of the inflationary psychology, other measures were being taken to assure a stable monetary and fiscal framework. Taxes were raised, and the Treasury and Federal Reserve System entered their famous accord, which permitted the Federal Reserve to continue a restrictive monetary policy. When controls were lifted in 1953, there was little postponed inflation.

HOW DICK WAS TRICKED

If the Korean War illustrates how to use controls, the experiment undertaken by the Nixon administration in August 1971 illustrates how *not* to use them. The evidence for an inflationary emergency at that time was weak; in fact, inflation appears to have been decelerating. Moreover, there was no attempt to use the time bought by controls to introduce a monetary and fiscal framework to assure long-run price stability. If anything, monetary policy became more expansionary after controls were introduced. Perhaps part of the reason was that the Federal Reserve and Congress assumed that they would no longer be blamed for inflation with the price authorities around to take the heat.

When controls were finally removed in 1974, there was a considerable wave of price increases. Much of this inflation, however, was the result of other shocks hitting the economy in 1974, and some of the inflationary pressure built up in the early phases of controls was dissipated in later phases. In retrospect, there appear to have been no long-run gains to justify the contortions the economy was put through in the early 1970s.

CONCLUSION

Would permanent controls work? Why not run the economy the way we did in World War II and enjoy permanently low inflation and high employment? Historical experience shows that, in general, permanent price controls would not work. The economy would be saddled with an ever-growing network of regulations designed to thwart ever-more devious evasions of the controls, reduced efficiency, and suppressed inflationary pressures likely to break out whenever controls were relaxed. What about a temporary use of controls, say for eighteen months or two years, to reduce inflationary expectations in an emergency? There is some evidence here that controls might work. The appropriate analogy is with a central bank fulfilling its role of lender of last resort. Normally, the rule should be let the market punish inefficient banks with failure. There are times, however, such as banking panics, when it is appropriate for the central bank to intervene, or so, at least, is the conventional wisdom. Something similar might be said about controls. Normally, the best rule is for the government to keep out and let the market decide the appropriate level of prices. There are times, however, such as "price panics" caused perhaps by the outbreak of war or a sudden shift to a more restrictive monetary policy, when direct intervention might work. The danger, of course, is that having admitted the theoretical possibility of a beneficial use of controls, politicians will use them for purely short-term, political reasons. The search for shortcuts and easy answers is thus the greatest danger to the rational use of controls.

18

The Securities
Exchange Commission:
Where From, Where To?

SUSAN M. PHILLIPS
and J. RICHARD ZECHER

Although the creation of the Securities Exchange Commission (SEC) is often attributed to the stock market crash of 1929, its establishment more properly should be seen as a response to the increased support for a regulatory program that promised to protect investors from manipulation and fraud. Over the years, SEC regulations have been modified only as a political response to economically influential groups rather than on economic grounds. Keeping this principle in mind, it is possible to predict the direction of future stock-market reforms with some degree of probability.

BLUE SKIES AND BUCKET SHOPS

If the nation needed a scapegoat for the stock market crash of 1929 and the ensuing Great Depression, Wall Street was an attractive candidate. Yet the creation of the SEC did not occur until nearly five years after the 1929 market crash, and then only after considerable political debate. Immediately after the crash, there were allegations of fraud and manipulation of the nation's exchanges, bringing calls for strong federal regulation of the securities markets. President Herbert Hoover, however, resisted such initiatives, claiming that the private institutions and the states should govern themselves. After all, the New York Stock Exchange (NYSE) had fairly elaborate self-regulatory structures dating back to the 1792 "Buttonwood Tree" agreement, which established fixed commissions and trading priorities. Furthermore, all states had "blue sky laws" that generally governed new issues of stock or bonds through state chartering of corporations to prevent the issuance of securities that were merely rights to purchase bits of blue sky. Many argued that these forms of limited regulation were not enough, but Hoover questioned whether federal regulation could even be authorized under the U.S. Constitution.

While the depth of the Depression kept pressure on government to "do something," there continued to be allegations of fraud on Wall Street. When the exchanges did not appear to be willing to exercise self-regulatory control over members to prevent abuses in exchange trading, Hoover appealed directly to Richard Whitney, head of the NYSE, to eliminate manipulation and to restrict activities on exchange floors. Whitney eloquently defended the freedom of the market place. In 1931, when the NYSE had taken no action in response to Hoover's request, Hoover initiated a Senate investigation.

The investigation continued through the 1932 presidential election and produced numerous headlines about security manipulations. A few brokerage operations had run so-called bucket-shop operations, passing money from one customer account to another without actually purchasing or selling securities for clients; some traders had spread false negative rumors or sold shares in the hope of driving stock prices down, thereby creating opportunities for investment at bargain prices; underwriting groups sold overvalued securities by overestimating companies' assets. Although these practices probably contributed to the initial stock market crash in 1929, it is unlikely that they were the cause of the stock market malaise that continued during the first part of the 1930s. It was only when the stock market remained depressed due to underlying economic weakness that securities legislation gained momentum.

Hoover took the tack that only specific wrongdoings should be punished, and continued to question the constitutional authority of securities regulation. The Senate's findings of fraud and abuse formed the foundation of a Democratic Party plank that called for federal regulation of holding companies, public utilities, and security and commodity exchanges. This made an interesting campaign issue, since Franklin D. Roosevelt, the Democratic candidate, was then governor of New York, the state in which many of the alleged abuses were occurring.

ROOSEVELT'S REFORMS

The Roosevelt New Deal and the mandate for securities regulation coming out of the 1932 election was clear. Roosevelt's first securities reform bill called for the regulation of securities by the Federal Trade Commission (FTC). The Securities Act of 1933, sometimes called the *truth-in-securities law,* has two basic purposes: (1) to provide investors with sufficient material information to enable informed investment, and (2) to prohibit fraud in connection with the sale of securities. The disclosure and antifraud pro-

visions of the 1933 Act established a precedent for subsequent securities-
related legislation.

Following passage of the 1933 Securities Act, Roosevelt continued his
quest of regulating the financial markets by appointing a committee, headed
by Secretary of Commerce Daniel C. Roper, to examine the need for a
regulatory structure in the stock and commodity markets. The Roper report
recommended a mild form of regulation including exchange licensing administered by a new agency staffed by stock exchange employees. The bills
introduced in the Senate and House of Representatives, however, were
considerably stronger, calling for more stringent regulation and oversight
of the exchanges by the FTC. Wall Street mounted a campaign to stop the
bills, but in light of the revelations of the Senate committee investigations,
few concessions were made to Wall Street in the Securities Exchange Act
of 1934. As a fall back position, the New York investment community had
argued for, and won, a separate agency over which they hoped they could
have some control. With the exception of its chairman, however, the composition of this new commission gave Wall Street little comfort. Roosevelt's
initial appointees included James M. Landis, George M. Matthews, and
Robert E. Healy, all from the FTC, Ferdinand Pecora from the Senate
committee, and as chairman and the only representative from the private
sector, Joseph Kennedy. Moreover, most of the securities division of the
FTC was transferred over to the new SEC.

The major requirements of the 1934 Securities Exchange Act included
registration of securities traded on national exchanges, certain periodic
financial reports of those securities, registration of exchanges and broker-
dealers, antifraud provisions, standards for transactions involving non-
public information, proxy and tender office solicitation, enforcement of
margin credit restrictions imposed by the Federal Reserve Board, and
prohibitions against "wash" sales, which were allegedly "no risk" sales
with repurchase agreements to create the illusion of volume and invest-
or interest. The Maloney Act of 1938 extended the 1934 Act to cover
over-the-counter securities and regulation of qualified broker–dealer
associations.

The early years of the SEC were marked by the building of the staff of
the agency, the launching of several major capital market studies, the
establishment of the various regulatory programs, and efforts by Chairman
Kennedy to restore public confidence in the markets through speeches and
other public relations efforts. The registration of exchanges resulted in the
closing of nine exchanges—including a one-person exchange in a poolroom
in Hammond, Indiana.

Under its first three chairmen, the SEC established itself as a reputable
and reasonable agency. James M. Landis succeeded Kennedy as chairman,
and William O. Douglas succeeded Landis. Before joining the SEC and
at the request of Chairman Kennedy, Douglas, a respected law professor
at Yale, headed a study of corporate reorganization that was the basis for
the Chapter IX Bankruptcy legislation. Under Chairman Douglas, there

were also significant reforms of the NYSE, perhaps hastened by the indictment of Richard Whitney, the leader of the old guard of the exchange, for misappropriation of securities belonging to the NYSE pension fund.

In short, while the SEC's origins surely can be traced to the market crash of 1929, the federal regulatory agency might not have come into existence except for the widespread dissatisfaction with Wall Street brought about by the Great Depression. Allegations of fraud and manipulation in the New York trading community further increased the support for the creation of federal securities and exchange regulation.

THE MAY-DAY TRANSFORMATION

After a number of years, during which it expanded its regulatory programs and jurisdiction, the SEC undertook its first and so far its only major deregulative initiative. On May 1, 1975, the NYSE permitted its members to set their own commission rates for the first time in history. Until that time, the NYSE had maintained a common fixed-rate structure for all public orders, despite some price competition from other exchanges both in New York and in other cities. Since the NYSE rules containing the fixed-rate structure were approved by the SEC, the SEC in essence had condoned the fixed-rate system. By the mid-1960s, however, the Justice Department was challenging the anticompetitive nature of fixed commission rates and urging the SEC either to justify or to eliminate them. That these political pressures were ultimately successful in introducing price competition to the NYSE is historically important, and raises an interesting question: Why, after 175 years, did such powerful opposition to price fixing arise suddenly in the 1960s?

To answer this question, it is necessary to understand the basic economics of brokering stocks. Briefly, as either the price per share increases or the number of shares per transaction increases, real commission costs change accordingly. The fixed commission schedule in place from 1959 to December 1968 reflected the price per share, but not the number of shares. After December 1968, the commission rate per transaction reflected the effects of both share price and number of shares. In neither case, however, did the fixed commission accurately reflect the actual brokerage costs incurred, as measured by the competitively set rates that emerged after May Day 1975.*

In comparing actual fixed commissions during the 1960s with estimates of competitive commission rates, two major patterns emerge. First, and

---

*This divergence between commissions paid and what would have been paid in a competitive environment is an estimate of the tax or subsidy on trades due to the fixed-rate commission structure, which affects a wealth-transfer between investor groups. An estimate of what competitive commission rates would have been during the 1960s can be provided by observing the post–1975 competitive rates and projecting them back in time. See Susan M. Phillips and J. Richard Zechner, *The SEC and the Public Interest* (Cambridge, MA: The MIT Press, 1981), pp. 64–90.

least surprising, fixed commissions were almost always set above the estimated competitive rates. Second, and more surprising, the difference between the fixed-rate charged and the estimated rate varied systematically with the share prices, and even more with the average number of shares per trade. In general, trades involving only 100 shares of an average-priced stock were charged a commission roughly equal to the competitive rates. As share prices increased, however, and particularly as the number of shares per trade increased, the fixed commission diverged more and more from the rapidly declining competitive commission rate.

In other words, the higher the average price of stocks traded, and the larger the number of shares per trade, the more the fixed commissions overcharged. For trades by individuals, it may be estimated that $35 million of the $136 million paid in commissions in 1960 exceeded the competitive rates. This amounts to what we may call a regulatory tax of about 26 percent for individuals in 1960. By 1968, largely because the average number of shares traded by individuals had almost doubled, this regulatory tax increased to $68 million.

For institutional investors, the difference between 1960 and 1968 was even more dramatic. The average number of shares per trade increased from about 200 to about 530. In 1960, institutions paid total commissions of about $85 million, approximately $55 million of which was above the competitive rates; by 1968, commissions paid had risen to $495 million, with $327 million representing charges above competitive rates.

Inasmuch as most of the excess was given back by institutions in various forms of non–price competition, these implied excess commission charges did not, in general, represent excess profits for brokerage firms. Nevertheless, non–price competition was inefficient, since it was very difficult to know which goods or services were of the most value to each customer without prices to set as signals.

The regulatory taxes paid by individuals and institutions peaked in 1968, when the NYSE implemented a new structure of fixed commissions that would cut excess payments roughly in half. When asked by the SEC to comment on the proposed new NYSE rate structure, the Justice Department became interested in fixed commissions as an anticompetitive practice. With public and congressional attention focused on commission rate fixing as a result of the Justice Department challenges and studies begun in 1968, it became increasingly difficult for the NYSE and its regulator to defend the exchange commission rate structure, which had, in fact, become increasingly anticompetitive throughout the 1960s. The SEC was eventually pressured to react, and the dramatic increase in the "regulatory taxes" paid between 1960 and 1968 reinforced this political response. In short, the regulatory tax to individual and institutional investors by the fixed commission rate structure became so great that the political response was to force deregulation, eventually resulting in the total dismantling of that rate structure. The Justice Department was a catalyst in the rate deregulation process, which was actually initiated by the NYSE itself in response to traders seeking to lower the cost of doing business by lowering the regulatory tax.

CONCLUSION

The basic approach to securities regulation over the years has not changed dramatically—full disclosure, antifraud regulation and enforcement, self-regulation. When one of the major tenets upon which the NYSE's self-regulatory programs, particularly the fixed-commission rate structure, began imposing heavy regulatory costs on investors as a result of changed trading circumstances, the SEC was forced to oversee a change to that structure. The lesson from this is that the SEC, its programs, and reforms are *political* responses to economically influential groups. Only if regulatory taxes become high enough is an SEC regulatory initiative or reform likely to be successful. Thus, although the 1987 stock market crash was significant, its effect on economically influential groups has not been great enough to prompt widespread securities reform. The SEC has proposed some far-reaching changes, but they have not been approved; instead, there have been several "fine tunings" of stock market regulations. Other events or factors, however, may have enough widespread economic impact to precipitate SEC reforms—for example, computerized trading systems or international competition. To the extent these developments unduly tax an economically influential, politically effective group of investors, changes to the SEC system of regulation are likely.

VI
TECHNOLOGY AND COMPETITIVENESS

In the sweat of your brow shall you eat bread. If you want more bread for less sweat, however, then technology must somehow get better. The government's role as referee in technological change is illustrated in the triad of tales by Gary Walton, Peter Temin, and Gary Libecap of lost monopoly. They illustrate the paradox of technological ownership, first emphasized half a century ago by Joseph Schumpeter. The paradox is that someone must own the patent on steamboats or the monopoly of telephones or the oil in the ground to make it worthwhile to innovate; however, monopolies on the other hand get comfortable and corrupt, and do not innovate. Ownership is necessary for good incentive, but if you own it, why bother?

The economist's praise for capitalism usually comes as praise for smart allocation. All the math runs this way, but capitalism's greater glory is innovation, not merely smart use of existing resources. Paul Uselding and Nathan Rosenberg stress the dynamic gains from capitalism, admitting a role for the government in setting the rules. Winston Churchill said once that democracy is the worst form of government . . . except all those other forms that have been tried from time to time. Similarly, although the math does not say so, history does: Capitalism is the worst form of economy . . . except for all the other forms that have been tried from time to time.

19

Fulton's Folly

GARY M. WALTON

Despite popular belief, it is not the free market that causes monopoly, nor the logic of laissez faire that constricts competition, but rather government intervention in the economy that gives unfair advantages to vested interests. Cable franchises, the Postal Service's exclusive government-operated first-class mail services, Amtrak's monopoly on passenger train routes, or telephone companies' lack of competition for local calls are not the products of *unfettered* capitalism. Established operators fight hard to eliminate competitive threats; corporate presidents line up with top union officials to press for legislative protection against foreign competition, while banks support statutes forbidding business across state lines. Of course, such monopoly privileges are typically asked for, and granted, in the name of the "public interest," but the common good is invariably better served by open markets, which generate productivity advances, stimulate trade, and lower the costs of goods and services. The role of government as a guarantor of monopoly, and the effects this has on the public interest, may be seen from many episodes in American economic history, but perhaps nowhere as clearly as in the colorful, early nineteenth-century era of steamboating.

THE GREAT STEAMBOATING MONOPOLY GRAB

Robert Fulton's dream was to introduce steam navigation throughout the civilized world. In 1811 he petitioned the Russian czar for steamboating monopoly rights between St. Petersburg and Kronstadt; one year later he signed an agreement with an Englishman to launch steamboating on the Ganges River in India. Soon he hoped to dominate water traffic within the United States. With one steamboat plying the Hudson River, and an absolute monopoly already granted him by New York for all steam navigation within the state, Fulton and his partner, Robert Livingston, schemed to win exclusive control over all steam navigation throughout the entire trans-Appalachian West, a natural highway network of approximately forty-five rivers extending over 16,000 miles. To lock up so vast an area, they diligently petitioned the main states and territorial assemblies along the trunk rivers, boasting of the benefits to the areas served, but claiming that their federal patent was of too short a duration to cover expected startup costs. Fulton and Livingston thus sought twenty years of exclusive control for building and operating one steamboat, plus an additional five years for a second boat and another five years for a third.

This petition was relentlessly fought by William Thornton, a spirited public servant who served as superintendent of the U.S. Patent Office. In a letter-writing campaign to the petitioned states and territories, Thornton pointed out that exclusive monopoly privileges based on state patents were expressly forbidden after 1789 by the Constitution of the United States. Proper jurisdiction between the federal and state governments on patents was still only loosely administered, however, and Thornton was trying to tighten up control as well as fight the monopoly plan. Thornton's campaign succeeded in most states, partly because of legal matters, and also because Western prejudice against "monied Easterners" provided natural sympathy to Thornton's admonishments. Kentucky's legislative committee rejected Fulton and Livingston's petition, warning: "It would be dangerous and impolitic to invest a man or set of men with the sole power of cramping, controlling, or directing the most considerable part of the commerce of the country for so great a period."

Fulton and Livingston, however, were not to be deterred, and they recognized there was really only one essential territory to control: the "Crescent City" of New Orleans, gateway to the outside world from the valleys of the Mississippi and Ohio rivers. Approval by the legislative assembly of the Territory of Orleans, known as the State of Louisiana after 1812, could give them the control they sought. One vital, *political* victory was all they needed.

To this end, a close friend of Fulton's came into contact with W. C. C. Clairborne, governor of the Territory of Orleans, and in 1810 invited him steamboating during his visit to New York. Like most Westerners, Clairborne had harbored doubts on the technical feasibility of steam power against mighty western currents, but such doubts were eliminated after several days of partying with with Fulton and Livingston. Clairborne duly agreed to sponsor their petition, and to personally present it to the legislative assembly, which granted exclusive steam navigation rights to Fulton and Livingston in April 1811. The monopoly was to last eighteen years.

The response was hostile throughout the West, especially in towns along the Ohio River. The Ohio and Kentucky legislatures passed resolutions denouncing the monopoly grant and urging congressional intervention. Mass meetings and broadsheet editorials called for "the annihilation of the swindling patent rights" that had allowed Fulton to lock the gateway to the West.

INTERLOPERS AND THE COURTS

Undaunted by protest, and with their hold on New Orleans secure, Fulton and Livingston launched a 371-ton steamboat, the *New Orleans,* at Pittsburgh in October 1811. It arrived in New Orleans in January, and successfully operated between there and Natchez, the most populous and profitable 300-mile stretch of river, until fatally striking a stump in 1814. In that same year, Fulton began operating the 340-ton *Vesuvius* and 360-ton *Aetna.* In 1815 he added another *New Orleans.*

Meanwhile, a group of "Westerners" near Pittsburgh defied the grant and quickly entered the business. Fulton and Livingston's most feared interloper was the mechanic and inventor Henry Shreve, whose vessels performed extremely well, thanks to his clever innovations and skillful adaptations to western river conditions. Legal skirmishes to halt Shreve and the other interlopers began in 1815, when one of Shreve's steamboats arrived in New Orleans; his vessel was seized and then released on bail, but other ships were impounded.

Fighting off Shreve's boats and other interlopers compounded Fulton and Livingston's practical difficulties. Particularly troublesome were the handling problems and power limitations of the monopolists' steamboats: Following designs more suitable to the Hudson River, their vessels sat too deeply in the water and were driven by engines too weak for the mighty Mississippi. By contrast, Shreve's boats were built of lighter wood with broad, shallow hulls for lighter draft, and were propelled by more powerful engines; to avoid snags and river debris, the paddle wheels were placed at the stern rather than the side. Of course, Fulton and Livingston eventually could have adopted Shreve's designs as did others. Their first order of business, however, was to win the court battle at New Orleans because as long as they had the Louisiana monopoly they could exclude superior competition and control the business to and from the interior. Despite an 1816 State Supreme Court ruling that the steamboat monopoly was unconstitutional, local politics remained in their favor. Although their exclusive privileges were detested throughout other western states, the Louisiana legislature rejected an 1817 petition to abolish their monopoly.

During these years of legal entanglements Fulton and Livingston died, but their heirs continued the court battle against Shreve. After the case was dismissed by the Louisiana State Supreme Court, Edward Livingston renewed legal proceedings in the U.S. District Court of Louisiana on behalf of his brothers' and Fulton's estates. Despite an alleged bribe to Shreve from Livingston to instruct his lawyers to throw the case, Shreve remained undeterred. The Federal District Court ruled that since both the plaintiffs and the defendant were citizens of states other than Louisiana, it had no jurisdiction over the case. Anticipating a fairly sure loss at the only tribunal left—the Supreme Court in Washington, D.C.—the heirs abandoned the fight.* Consequently, open competition on the western rivers was assured

---

*Indeed, in 1824 the Supreme Court ruled that their New York exclusive steamboat franchise

by the end of 1817, when sixteen steamboats were running between Louisville and New Orleans; two years later there were fifty-nine steamboats plying the trunk rivers.

COMPETITION ON THE RIVERS

The gain to the West and the entire nation from free entry and open competition on the rivers was immense. Because the steamboat monopoly was broken, freight rates fell faster and reached lower levels than would have prevailed otherwise. Also, on account of competition, freight shipments grew faster and reached higher levels, and steamboats contributed mightily to the early nineteenth-century transportation revolution. If Fulton's monopoly had been protected, the West would have been populated and developed, of course, but at a slower pace; the impact of another dozen years of monopoly at New Orleans would not have been trivial. It is difficult to imagine that any court or legislative body could have withheld the powerful forces encouraging vessels to enter these lucrative markets for long.

Huge profits were at stake. Fulton's *New Orleans* cleared $20,000 in its first year (1811) over and above expenses, repairs, and interest. The investment in the vessel had been about $40,000. Even at midcentury, with railroads coming into their own, river transport still played a vital (and profitable) role. In 1849, western river steamboats logged 3.3 billion freight-ton miles and 1.1 billion passenger miles—1 billion ton-miles more than all of the nation's railroads were hauling ten years later.

Falling costs of transport and an open gate for trade and commerce encouraged rapid settlement in the western lands. Total outbound freight tonnage from the country's interior growth grew from 65,000 tons in 1810 to 7,690,000 tons in 1860. Over these same fifty years, the population of the trans-Appalachian West grew from 1 million to 13 million people— from 15 to 41 percent of the total population. Freight traffic out of the vast midwestern interior through New Orleans accounted for over 80 percent of all outward shipments until the late 1830s. This southern gateway was not surpassed by the *combined* northern (Great Lakes, St. Lawrence, Erie Canal, Hudson River) and northeastern (canals and railroads) gateway routes until the early 1850s.

The dominance of the southern gateway throughout this period of rapid western expansion attests positively to Fulton's vision of enormous profits from controlling the lock on the steamboat's gate. That control never became reality, however, and the fortunes that might have been Fulton

was an unconstitutional violation of the interstate commerce clause. According to Archibald Cox, this 1824 steamboat monopoly case ruling was one of the most important constitutional decisions in U.S. history: Archibald Cox, *The Court and the Constitution* (Boston: Houghton Mifflin, 1987) as cited in Thomas W. Hazlett "Duopolistic Competition in CATV: Implications for Public Policy," *Yale Journal on Regulation* 7 (January, 1940), fn. 49.

and Livingston's were consequently more widely shared. So, too, were the gains generated by many competitors, who scurried to find better, less costly ways of operating steamboats. Advances in steamboating productivity, propelled through competitive markets, forced the real costs of shipment downward at an unprecedented pace. Freight rates, in 1820 constant dollars per 100 pounds shipped upstream from New Orleans to Louisville, fell from $3.12 to $2.00 between 1810–1815 and 1820. They then fell to $0.28 by 1860. Shreve and other noted innovators helped realize these later improvements, but major credit must go to many, mostly unknown, contributors. A host of minor improvements, primarily from learning-by-doing, led to fundamental changes in the basic physical characteristics, handling, and operations of the vessels. The ratio of carrying capacity to vessel size increased threefold, while river improvements and clearing for tree trunks and obstacles lengthened the boats' useful lives. Although running time speeds changed little, the average number of yearly round trips per vessel quadrupled between 1810 and 1860. These gains came from reduced port times and from increases in the number of months each year vessels could safely navigate the rivers. In the words of historian Louis C. Hunter:

> The story is not, for the most part, one enlivened by great feats of creative genius, by startling inventions or revolutionary ideas. Rather it is one of plodding progress in which invention in the formal sense counted far less than a multitude of minor improvements, adjustments, and adaptations. The heroes of the piece were not so much such men as Watt, Nasmyth, and Maudslay, Fulton, Evans, and Shreve—although the role of such men was important—but the anonymous and unheroic craftsmen, shop foremen, and master mechanics in whose hands rested the daily job of making things go and making them go a little better.

Tight monopoly control would have thwarted these many minor advances that, collectively, caused the decline in water transportation costs and opened the agricultural Midwest to feed a hungry world.

THE CASE FOR OPEN COMPETITION

History students and enthusiasts will find at least two familiar lessons in the story of Fulton's folly. First is the observation that it is very difficult to establish or sustain a monopoly without government patents, licensing, tariffs, or regulation to keep out competition. Once Fulton and Livingston lost legal protection they were overwhelmed by competitive forces, and no single owner or group of owners could stem the tide of new entrants who were soon beating down freight rates. A multitude of suppliers responded, as if led by an invisible hand, to build and operate more and more boats at lower and lower costs. No central planning was involved here, just a wide-open competitive fray.

The second lesson is that open markets generate a multitude of productivity advances, large and small, that stimulate trade and lower the costs of goods and services. Because of competition in the steamboat business, consumers received more goods at lower retail prices, and producers sold their farm goods and other merchandise in greater volume at higher wholesale prices. As freight rates tumbled, the gap between retail and wholesale prices fell. The only notable losers were the handful of keelboat owners who were swiftly driven out of business on the trunk rivers. Even so, many of these old boats extended their working lives by retreating to the remote tributaries until the steamboat eventually arrived there, too. Men hired to pull and pole keelboats upstream lost backbreaking jobs, but readily found easier work on steamboat crews. Moreover, one old technology, the flatboat, was actually boosted by the new technology. Steamboats aided flatboatmen by providing comfortable upriver return transport at a fraction of the costs of doing it the old fashioned way, by foot.*

CONCLUSION

We have seen that monopolies are difficult, if not impossible, to sustain without government sanction. We have also seen that competition produces a variety of advances in productivity that monopolies, by their nature, cannot provide. Thanks to the U.S. Constitution, state-chartered monopolies of the type envisioned by Fulton and Livingston have seldom succeeded for long in the United States. American's distaste for monopoly has a long and historically rich tradition; the American Revolution was sharply spurred by colonial reactions to the granting of a monopoly in tea to the English East India Company in 1773. Nevertheless, the quest for monopoly is an eternal force that begs for heroes to fight against it and sustain the progress of competition. The heroes in Fulton's Folly are William Thornton, Henry Shreve, and all the little guys who added to progress through competition. Each generation, it is hoped, will energize likeminded citizens to combat the temptations of exclusive control.

20

Down the Primrose Path

PETER TEMIN

Bernie Strassburg was frustrated. Trained as a lawyer and initiated into the world of public policy in the heady trust-busting days of the New Deal, he had risen through the ranks of the staff at the Federal Communications Commission (FCC). The year was 1968, and Strassburg had been chief of the Commission's Common Carrier Bureau for five years. His job was to regulate telephones and telegraphs—which meant, for all practical purposes, trying to regulate American Telephone and Telegraph (AT&T), the largest corporation in the world. AT&T monopolized the markets for local and long-distance telephone service, as well as the markets for telephone equipment, because the federal government made it illegal for anyone else to enter these markets. Strassburg was supposed to regulate the monopoly in the name of the public interest, but as he moved to get control over AT&T, he felt himself to be a Lilliputian trying to tie down a modern corporate Gulliver. At the time, he had no idea that his actions would set the stage for the dissolution of the Bell System as a whole, but this, in fact, would be the result. The breakup of AT&T's telephone monopoly was caused by Bernie Strassburg's attempt to get control over the regulatory process, and by the faulty economic analysis that informed his attempt.

ROUSING THE GIANT

For more than a decade, Strassburg had felt that AT&T was unregulatable. In the mid-1950s, when the phone company had been accused of violating the Sherman Antitrust Law by monopolizing the market for telephone equipment, AT&T was unwilling to even consider the possibility of restructuring the American phone system along European lines, with separate operating and manufacturing companies. After all, the company's leaders saw no reason to give up the their Bell System, which provided local telephone service through the Bell Operating Companies, long-distance telephone service through AT&T's Long Lines, and telephone equipment from Western Electric and Bell Labs. When the Eisenhower administration had gone along with the phone company and asked the FCC if it could regulate the integrated Bell System, Strassburg drafted the Commission's reply, warning that while the FCC had the legal authority to regulate AT&T, it lacked the resources and expertise to control the giant Bell System.

The letter that was sent to the Justice Department retained the passages about legal authority, but Strassburg's all-important qualifications about their effectiveness had been excised by some obsequious higher-up. Strassburg's attempt to reveal that even then AT&T had become too large to regulate had failed. Citing the FCC's letter as support, the Justice Department negotiated a Consent Decree with AT&T in 1956 that protected the Bell System.

Then, in the late 1960s, a new set of challenges to the FCC had arisen as the expanding demand for private communication, as opposed to public broadcasting, was matched by the progress of the technology that enabled supply to meet demand. Sensing that the age was soon to dawn when data processing and telecommunications would become interconnected, Strassburg got ahead of the curve by initiating a "Computer Inquiry" to set FCC policy for the new technological era. In so doing, Strassburg presided over the abolition of the Bell System's restrictions on the connection of customer-owned equipment to the telephone network, opening the way to the independent supply of telephones, answering machines, and modems. Now, the FCC was considering an application from a tiny company called MCI to furnish what the company termed a new service.

Telephone messages traditionally had traveled along wires, as they still do from your house, but engineers had learned during World War II how to send signals over the air by the use of microwave radio, which could be directed and focused like light. The FCC had decided in 1959 that the Bell System did not have exclusive rights to the use of microwave radio, allowing other firms to build microwave systems for their own use; MCI proposed to use microwave radio to offer a new kind of private line service between St. Louis and Chicago. The company wanted to build a microwave facility that would be rented to other companies' private-line services on that route. The users would no longer have to build their own facility to have independent private lines, at least on this one route; they could rent from MCI. Instead of facing a choice between using the Bell System or building their own facility, they would face a choice of existing facilities to rent.

It would be a useful choice because MCI's proposed new service was to be a markedly cheaper version of AT&T's, with roughly the same quality. Consumers, particularly the executives of business firms large enough to use private-line communications, always love a bargain, and MCI had no trouble finding people who said they had a "need" for the new service MCI would offer. They were really saying that they were happy to move along their "demand curve" (i.e, that they would buy more of a service or have a greater demand for it) if the price was lower. The FCC, however, being unschooled in the shape of demand curves, did not appreciate the role of price in their answers: The Commission could not distinguish between a "need" for a new service and a desire to purchase more of an existing service at a lower price.

MCI claimed that it could offer cheaper service because it was more efficient than the Bell System in the use of the new microwave radio technology. The truth, however, was quite different. AT&T's *costs* probably were not very much different from MCI's. Its *prices* for private-line services, however, were very much higher than MCI's. These prices were set by AT&T under the terms of the Federal Communications Act of 1934 as interpreted by the FCC. They were regulatory prices, set at a time when there was no competition for the Bell System's services. Since MCI could charge lower prices than the FCC would allow the Bell System to charge, MCI asked the FCC for permission to do so. Potential customers said they would like to use MCI's new services. AT&T replied, with some justification, that MCI was just going to skim the cream off of the Bell System's regulatory prices.

Strassburg had little difficulty reaching a decision. Already searching for ways to control AT&T, he wanted to introduce new technologies faster than the Bell System appeared willing to move; he needed a pin to prick the giant telephone company, and MCI looked like the answer to his prayers. It was a tiny firm, with less than 100 employees and no revenues at all. It was proposing to offer a limited service on a single route. It could not possibly hurt AT&T, but it might rouse the sleepy giant to action and to the need to move faster to keep up with a rapidly changing world. Strassburg supported MCI's application, contending that it would be an experiment in providing varied services with very little risk of harm. Not everyone at the FCC agreed; some foresaw dangers to the integrated telephone system, or doubted that MCI had the financial or technical strength to deliver on its promises. Strassburg carried the day, but barely: MCI's application was approved by the FCC in 1969 on a four-to-three vote, with the Commission's chairman dissenting. With only a fraction less support, MCI would never have gotten off the ground. Still, the battle was won— and yet, it had only begun.

GOODBYE TO BELL

The 1969 MCI decision, far from being a harmless experiment, started American telecommunications along the path that led to the end of the Bell System fifteen years later. The subsequent history was not predetermined by the FCC's decision, but the direction of public policy became ever harder to change. Strassburg clearly did not anticipate the consequences of this single decision. He had taken an important step down the primrose path without realizing that he was on it.

Failing to see that MCI was simply skimming the cream off of the Bell System's regulatory prices, Strassburg could not see the dynamic set up by his decision. Each time the FCC allowed firms to enter and take advantage of gaps between competitive and regulatory prices, the entering firms made profits from the price spread and then used their profits—or sometimes the expectation of profits—to push the FCC for more opportunity. Each decision allowing more competition in telecommunications intensified the pressure for even more competition. The FCC would find that it could not take one step down this path without taking the next.

The immediate results of the decision on MCI demonstrated the process. John McGowan, the aggressive venture capitalist who had taken over MCI while its application was pending at the FCC, was not interested in Strassburg's experiment. He did not have the time to build a microwave facility on a single route and watch for its effects. The profits available on the St. Louis to Chicago route were available on other heavily utilized routes where the Bell System's regulatory prices were far above the costs of an independent facility. He wanted to capture those opportunities before others came in. Instead of concentrating on this single route, therefore, McGowan set up MCI-affiliate firms all over the country and flooded the FCC with applications to build microwave facilities on other routes. Within a year, the FCC faced approximately 2,000 applications for private microwave facilities from various MCI companies as well as from another aspiring entrant called Datran.

Strassburg was overwhelmed. There was no way to process all these applications on the individual basis used for MCI's initial project. He consequently switched from considering them individually to making a rule for all of them at once. This strategy changed the legal framework for the decision and obviated the need for public hearings. The FCC issued its *Specialized Common Carriers* decision in 1971, allowing independent microwave facilities to be constructed under very general circumstances.

In two short years, Strassburg's "experiment" had been converted into public policy. There had been no time to assess the implications of the initial decision, nor time to get feedback on its effects. No new services were offered, nor was any technology not already exploited by the Bell System introduced. Instead, McGowan had been unleashed to push ever harder to get a piece of the Bell System's national network.

His next effort was to obtain connections between his facilities and the Bell System. As long as private microwave facilities had been used by single firms, no connections had been needed; however, MCI proposed to rent capacity to different firms. It did not want to build physical connections to each renter; it wanted to make its service as much like the Bell System's as possible, even using the Bell System's facilities to do so, while charging competitive prices and confining its attention to those routes where competitive prices were below regulatory ones.

AT&T only reluctantly complied with MCI's interconnection requests. In its view, MCI was proposing to "piece out" the lucrative parts of the national network, using its freedom to charge competitive rather than regulatory prices to steal business from the Bell System. AT&T responded grudgingly to the regulatory and judicial decisions on interconnection.

MCI, impatient with the treatment it was receiving, filed an antitrust suit against AT&T and worked to convince the Justice Department to do the same. In 1974, less than twenty years after the 1956 Consent Decree, AT&T was again accused of violating the Sherman Antitrust laws. It was, of course, the agreement reached to settle the government's suit that broke up the phone company.

MCI also offered a new service in 1974. Called "Execunet," the new service utilized MCI's interconnection with the Bell System to offer switched long-distance service to subscribing customers. The rates were far lower than were AT&T's because MCI's service was not subject to the separations process; MCI subscribed to local Bell Operating Companies at ordinary business rates. By contrast, AT&T paid much more for its interconnection under the separations rules that declared that its customers contributed to the upkeep of the local plant each time they made a long-distance call. Indeed, MCI's Execunet service was a recipient of the separation charges imposed on AT&T's message-toll service.

The FCC, urged on by AT&T, objected to MCI's new service, and the controversy extended into the courts. The FCC argued that its *Specialized Common Carriers* decision had allowed the construction of microwave systems for private lines only. MCI replied that its Execunet service was a kind of temporary private line and, even if it was not, the facility whose construction was approved by the FCC could be used for any service.

The Federal Appeals Court agreed with the last of MCI's arguments in 1977. Execunet was an ordinary long-distance service, not a private-line service, but the FCC could not bar MCI from offering it without full evidenciary hearings. Since the FCC had not even considered switched long-distance service in its deliberations over *Specialized Common Carriers,* it could not bar MCI from offering it. Strassburg's experiment—a private line facility on a single route—had turned into competition in switched long-distance service in only a few short years.

By the time the government's antitrust suit against AT&T came to trial in 1981, therefore, it was in the context of competitive telecommunications. The lawsuit, initiated at a time when competition was restricted to corners of the telecommunications market, was settled after competition had spread throughout the system. In order to keep its vertical integration—an aim of AT&T's management in 1981 no less than in 1956—AT&T was forced to give up its ownership of the Bell Operating Companies. The Bell System ceased to exist.

CONCLUSION

At no point in the process initiated by Bernie Strassburg was there serious, dispassionate consideration of the difference between regulatory and competitive prices. Without understanding how historical, average-cost pricing differs from marginal-cost pricing, AT&T's regulators could not understand the cumulative nature of the process they were in. The regulators and judges consequently did not see that the "new" services being offered by potential competitors were only movements down along the demand curve for existing services. They did not understand that incentives for entry came from profit opportunities resulting from different pricing methods. They also did not realize that each time they allowed one group of entrants to shelter under the Bell System's price umbrella, they encouraged others to follow. Whether the result was good or bad, the process by which the Bell System was dismembered was a triumph of inadequate economic analysis.

21

What Really Happened
at Teapot Dome?

GARY LIBECAP

In 1921, petroleum executives E. L. Doheny and Henry Sinclair allegedly bribed Secretary of the Interior Albert Fall to obtain oil production leases for the Naval Oil Reserves at Elk Hills, California, and Teapot Dome, Wyoming. This "Teapot Dome" scandal, as it came to be called, is the most frequently cited example of the corruption of the Harding administration, and its legacy remains for many one of the darkest chapters of U.S. public lands policy. There is, however, another, possibly more important legacy of Teapot Dome. Surprisingly, Teapot Dome may have more to say about the success of future U.S. oil policy than it does for evaluating the ethics of government officials. After all, Teapot Dome was about oil, and the reasons it was so controversial, aside from the subsequent issue of bribery, tell us something about how we manage or, rather, mismanage our precious domestic oil reserves today.

THE COMMON-POOL PROBLEM

After the Spanish–American War of 1898, the U.S. Navy began converting its ships from coal to oil power. Other navies were undergoing similar conversions, and U.S. naval leaders believed that American naval power depended on a move to a more versatile fuel. The navy had petroleum deposits on government lands. Beginning in 1912, naval oil reserves were created at Elk Hills and Buena Vista Hills in California, and, in 1915, at Teapot Dome in Wyoming.

These reserves were to provide a long-term source of supply of oil for the navy, but as with all U.S. oil reservoirs, they were threatened by serious losses from so-called common pool production. Ownership was assigned only upon extraction; until then, anyone could claim the oil.

The practice, then as now, was for land owners, including the federal government, to grant oil production rights through leases to firms on both federal and private lands. Because reservoirs were often large, while the leases were small, many firms found themselves competing for the same oil. A firm could drain its neighbor's leases by drilling multiple wells and rapidly pumping the oil. The ensuing race for petroleum led to the drilling of many more wells than would be needed merely to extract the oil. As the oil was removed from the reservoir by each firm, it was stored in extensive and costly surface storage facilities that would not otherwise be needed if the oil were withdrawn more slowly. As firms pumped feverishly to extract the oil before their neighbors could, they vented the underlying natural gases needed to propel the oil to the surface. When subsurface pressures fell, due to rapid drilling and pumping, oil became trapped below the surface and recoverable only at very high extraction costs.

The costs of common pool production in the early twentieth century were indeed quite high. In 1910, estimates of oil losses from fire and evaporation in California alone ranged from four to eight million barrels, which was 5–11 percent of the state's production. In 1914, the director of the Bureau of Mines estimated losses from excessive drilling in the United States at $50 million, when the value of U.S. production was $214 million. These losses of common pool production threatened the viability of the naval oil reserves.

One problem was that there were many small leases on private lands within the naval reserves, and the firms that held those leases were drilling as rapidly as possible to extract the navy's oil. Of the 30,080 acres of the Buena Vista Hills reserve, 20,320 acres were private land in 640–acre plots, checkerboarded across the reserve because of past railroad and state school grants of federal land. Of the remaining 9,760 acres, 7,520 were contested by private claimants. Hence, there was little land in Buena Vista Hills that could be considered a true naval reserve.*

The Elk Hills and Teapot Dome reserves were less fragmented by private holdings, yet they too were vulnerable to common pool losses from rapid production on small leases on federal land within or adjacent to the reserves. Under the Mineral Leasing Act of 1920, which applied to federal lands not in the reserves, the secretary of the interior could grant leases to private firms for up to 2,560 acres for prospecting and for 640 acres for production. When reservoirs often covered thousands of acres, however, this leasing policy resulted in many firms competing for the same oil. Under a related law, the secretary of the navy could issue similar leases to private

---

*Elk Hills and Teapot Dome were more viable storage sites because they had fewer private leases. The Elk Hills reserve of 37,760 acres had less checkerboarded land because of a U.S. Supreme Court ruling (251 U.S. 115) cancelling much of the Southern Pacific railroad land grant in the area. Nevertheless, there were 6,760 acres of private patented land, checkerboarded in the reserve, and other private claims made through various U.S. land laws. Within Teapot Dome there was no patented land, although one company, Pioneer Oil, claimed all but 400 acres of the 9,481-acre reserve.

firms to drill within the naval reserves. Both policies set the stage for relatively small leases and competitive, common pool production.

DRILLING FOR SCANDAL

This was the condition that faced Interior Secretary Fall in 1921. Elk Hills was being drained by Standard Oil production from fifty wells on two sections within the reserve, and from twenty-nine wells directly along the reserve's boundary. Additionally, private wells operated by other companies were scattered within Elk Hills. Buena Vista Hills, more seriously checkerboarded, was subject to drainage from 599 wells on private land within the reserve, and from 183 wells on adjacent lands.* Teapot Dome in Wyoming was similarly subjected to drainage and other reservoir damage due to production on adjacent Salt Creek field, though development was not yet at the level in California.

To help reduce oil drainage from its lands, the navy authorized small leases for drilling offset wells just inside the borders of Buena Vista and Elk Hills in 1920. The leases, however, were not assigned until 1921, when the Harding administration took office, and they could not prevent the decline in reservoir pressure. In July 1921, Secretary of the Navy Edwin Denby transferred jurisdiction over the naval reserves to the Interior Department to implement new policies for naval oil supplies.†

Interior Secretary Fall could have granted numerous, small 640-acre leases on Elk Hills and Teapot Dome, similar to those authorized by the Mineral Leasing Act of 1920 for other federal lands, but instead he issued single-firm leases to virtually all the federal acreage in Elk Hills and Teapot Dome. Production rights to over 30,000 acres in Elk Hills were granted to Pan American Oil. In exchange for the leases, the company agreed to pay a royalty to the federal government, and to construct storage tanks at various naval reserves. Similarly, Fall issued a single lease for the Teapot Dome's 9,481 acres to one firm, Mammoth Oil. Under the lease terms, Mammoth agreed to pay a royalty of 12–50 percent, to construct a pipeline to Kansas City, and to exchange royalty crude oil for fuel oil on the Gulf and Atlantic Coasts.

These large leases offered considerable advantages. The incentive to competitively drill and drain was eliminated; with only one firm on the reservoir, there was no need to sink wells to drain neighboring leases. Both

---

*The Interior Department estimated that approximately 22 million barrels had been drained from the two California reserves by March 1921, with the government's royalty share of the lost oil equal to 6.8 million barrels. More seriously, the Department concluded that private production on neighboring lands was so depleting subsurface pressures within the reserves that total oil recovery was being reduced by approximately 24 percent.
†The most fragmented reserve, Buena Vista Hills, was subsequently leased to multiple firms by Interior Secretary Fall and was no longer considered a reserve for the navy. Elk Hills and Teapot Dome, the remaining naval reserves, were to be leased to private firms with the oil placed in surface storage tanks in strategic locations for use by the navy.

Pan American and Mammoth Oil could adopt longer-term production practices that would reduce the number of wells drilled and the amount of surface storage tanks built. More important, the companies could adopt more measured production strategies to avoid the premature venting of subsurface pressures, maximizing overall oil recovery from the reservoir. Indeed, Fall's leases on Elk Hills and Teapot Dome were the first examples of unitized or noncompetitive oil production as a means of reducing waste in the American oil industry.

The large, single-firm leases were immediately challenged, however, by conservation groups and by small, independent oil companies in Wyoming that were excluded from Teapot Dome. Small companies were further incensed by the large leases when Secretary of the Navy Denby ordered the marines to eject an independent producer, Mutual Oil, from Teapot Dome for trespassing; conservationists, for their part, objected to any oil leases, large or small. These two groups pressured Congress to review Interior policies, and the hearings lasted from 1922 through 1928.

Despite intense opposition, the naval oil reserve leases likely would have remained intact had Fall not been convicted for receipt of $100,000 from E. L. Doheny and Henry Sinclair. Although neither Doheny nor Sinclair were themselves found guilty, the corruption issue doomed the leases, which were cancelled in 1927, and the reserves were transferred back to the navy for administration. As a result, the leases were not in effect long enough for the benefits of unitized production to be demonstrated.

In 1938, the Navy obtained authorization from Congress to consolidate lands within the reserve and, again, to lease it to a single firm, Standard Oil. There was no public outcry this time; the political battles over administration of other federal lands had been settled, and Congress was paying much less attention to oil policy. Meanwhile, provisions for large-lease, unitized oil production had been incorporated into law for all federal lands through the 1930 amendment to the Mineral Leasing Act of 1920. That amendment strongly encouraged, if not mandated, unitized oil production on federal lands. Interior Secretary Fall seems to have been farsighted in his 1921 leasing policies, and the benefits of unitized production were incorporated in all federal leases after 1930. Due to opposition from small leaseowners, however, the progress of unitized production has been much slower on private lands.

THE UNITIZATION SOLUTION

Many of the common-pool losses so clearly outlined in the early 1900s continue today. Because of the incentive to competitively drill and drain, U.S. firms drill more wells than would otherwise be necessary to pump oil. In 1980, the United States had 88 percent of the world's oil wells, but only 14 percent of the world's oil. In addition, more than 50 percent of the oil in any reservoir is never extracted because of the early loss of subsurface pressure. Vast quantities of potential domestic oil are thus lost because of common-pool production.

The problem of small-lease production is even more severe on private land than it is on federal lands. With fragmented surface ownership, hundreds of firms may have production leases to a large reservoir with each firm having an incentive to drill and drain. For example, on West Texas' Hendrick field in 1928, competitive drilling had led to the construction of storage tanks with a capacity of 11 million barrels at a cost of $3.8 million, while on the neighboring and larger Yates field, where there were larger leases and fewer firms, storage existed for only 783,000 barrels at a much smaller cost of $274,000. These and other common-pool losses could be avoided by unitization, under which all the small leases on an oil field would be combined and one firm would be selected by leaseholders to develop the field. The leaseholders would then share the costs and returns of oil production with other parties, including the firms that would otherwise be producing on the field. With a single-producing firm on a reservoir, the incentive to competitively drill and drain is removed.

The key barrier to the unitization of any given reservoir is the holdout of small lease holders. Firms with small leases are more likely to conclude that their share of the net revenues within the unit is less than what they would receive under either open production or state regulation.

One answer to the holdout problem by small lease owners would be compulsory unitization. Efforts to pass compulsory unitization legislation in Oklahoma and Texas began in the 1930s, but were successfully blocked by intense opposition from small firms. A compulsory unitization bill was not passed in Oklahoma until 1947, and none has been enacted in Texas. Even in Oklahoma, the legislation is written in such a way as to delay unitization; the field must be fully developed before a unit will be approved by the state. Despite government attempts at coercion, unitization remains a solution to the common-pool problem that has not been fully tried.

CONCLUSION

The historical case of Teapot Dome teaches valuable lessons about how we mismanage our precious domestic oil reserves. Specifically, we have yet to come to terms with the common-pool and small-lease problems. Currently, there is political pressure to place a tax on oil imports to encourage domestic exploration and production. History tells us, however, that in most of the major producing states, domestic oil will be produced in very costly ways that leave much of the oil trapped below the surface. To help avoid that dismal prospect, Congress could require, as a condition for an import tax, that compulsory unitization laws be passed in each of the major producing states to encourage early unitized exploration and production. In that way, the small-lease problem in production could be avoided and more new oil produced at a lower cost. Successful federal government pressure on the states for compulsory unitization on private lands could thus be one positive legacy of Teapot Dome.

22

Does Government Intervention in the Economy Hurt Competitiveness — Or Help It?

PAUL USELDING

Although it is widely believed that the rapid pace of American technological advance has been due to the incentives inherent in a system of laissez-faire, the federal government was a catalyst for technological advance during the nineteenth and for much of the twentieth century. For instance, federal legislation and support was instrumental in the development of precision machining industries, in lifting agricultural productivity, and in the genesis of computer technology. Indeed, the Federal government has encouraged innovations and their diffusion throughout the private market economy throughout most of our history—a fact of special relevance to the current policy debate about national economic performance or competitiveness.

THE GENIUS OF JEFFERSON

The technology of mass production uses specialized machines to abridge manual labor and produce large volumes of standardized products at high quality and low unit cost. The relative abundance and inexpensiveness of automobiles, agricultural machinery, and many kinds of consumer durables that characterize our national affluence is attributable to the technology of interchangeable manufacture and mass production. This technology was created in the American firearms industry during the first half of the nineteenth century, from where it spread to every industrial sector involved in the manufacture of producer and consumer durables.

At the beginning of the nineteenth century, the firearms industry was comprised of a few gunsmiths possessing considerable skill and resourcefulness, but who made each piece individually. As far as the nascent government was concerned, there were two problems with this means of production: the total number of arms produced by hand-crafted methods was far short of the military needs of the country, and such arms as were produced by gunsmiths could not be easily repaired in the field in the event of damage to one of the component parts.

Since Britain still had designs on her lost colony, and since French military needs were great because of hostilities with Britain, the ordinance deficit could not be filled by importations from abroad. With the capacity of the U.S. domestic industry severely limited relative to national needs, a solution had to be found. As early as 1793, Thomas Jefferson had been instrumental in securing an arms contract for Eli Whitney. Whitney was obligated to produce 10,000 stand of arms in two years, and, further, they were to be made of fully interchangeable components. The notion of interchangeability was one that Jefferson had picked up during his diplomatic assignments in France. Whitney, the celebrated inventor of the cotton gin, was seen as just the person to make this grand concept a reality; however, as talented and creative a mechanic as there was at this time, Whitney failed in his quest.

The solution to the problem was found in two actions of the Ordinance Department, then organized under the secretary of war. In the early 1790s President George Washington, drawing upon his knowledge as a military leader, established two federal armories to store, repair, and produce limited quantities of arms to ensure that there would be a secure supply of ordinance in time of need. Shortly after 1812 the chief of ordinance reorganized the federal armories, moving the system of production away from organized gunsmithing to something reflecting modern manufacturing practice that was characterized by the use of special-purpose machinery and division of labor.

Underlying the reorganization was the requirement that all piecework was to be manufactured according to the "Uniformity Principle." To insure that the principle of uniformity was adhered to, skilled mechanics were stationed at each armory. In addition, the Ordinance Department contracted with private gunmakers to supplement production at the federal armories. The inspectors checked the work of the private contractors, but, more important, served as a diffusion mechanism. As they traveled from one armory to another, they reported on new production methods and systematically oversaw their adoption in the federal armories. By the 1830s, the national armories became state-of-the-art facilities that had achieved practical interchangeable manufacture, the forerunner of modern mass production. As mechanics at the national armories moved to employment in other industries, such as clocks, sewing machines, or agricultural machinery, they carried with them knowledge of this new method of production. These achievements were possible because of the way the Ordinance Department organized the arms procurement function with a system of mixed public and private production; the steady patronage of the private contractors permitted investment in fixed capital at reduced risk; and adherence to the Uniformity Principle.

It is worth noting that the Ordinance Department acted in a facilitative and supportive fashion and was careful to provide incentives to the private arms contractors, while being sure to integrate the contractors' ideas and innovations into their own facilities at the national armories.

FORMING A MARKET NEXUS

Federal involvement in establishing agricultural experiment stations began with the passage of the Hatch Act in 1887. Prior to that time, there had been a long and heated conflict between proponents of practical training for farmers, and those who wished to give agriculturalists a broad science background modeled upon the laboratory methods of leading European universities. American agricultural practice up to this time can only be described as wasteful and land-intensive, promoted by the careless and prodigal alienation of the public lands, culminating in the Homestead Act of 1862. Agricultural practices lacked both an undergirding body of systematic knowledge and teachers capable of imparting it.

In 1862 the Morrill Act, granting land endowments to state colleges offering courses in agriculture and mechanic arts, signaled the beginning of a process by which a nucleus of teachers and research workers were recruited from native scientists educated in old-line colleges and medical schools. The agricultural experiment stations were a means of focusing this developing body of talent on the problems of producing better varieties, breeds, methods, and techniques. The range of work in these stations covered botany, soils, fertilizers, crops, horticulture, entomology, feedstuffs, animal nutrition, and dairying.

The combined effects of the land-grant colleges and agricultural experiment stations provided decentralized, science-based agricultural laboratories throughout every principal farming region in the United States. Coupled with the extension services of the land-grant colleges and universities, which served as a critical link in the diffusion of agricultural innovations, the experiment stations were instrumental in lifting agricultural productivity. The key to this outcome was the facilitative and supportive role of the federal legislation, which provided the initial resources for the establishment of the land grant system and the annual operating funds for the experiment stations. University control, decentralization, and the emphasis on delivery and diffusion of new knowledge allowed the market nexus to be formed. This connection between government support and patronage producing new methods, practices, and information in state-of-the-art facilities, open to private market interaction, worked as effectively in agriculture in the late nineteenth century as it did in the precision machining industries earlier on.

CREATING A COMPUTER

While there are many antecedents to the modern digital computer, the main line of its developments is generally attributed to J. Presper Eckert and John Mauchly at the Moore School of Electrical Engineering of the University of Pennsylvania. In 1942 members of the Moore School faculty, in conjunction with the U.S. Army's Aberdeen Proving Ground, were computing artillery firing tables. The computations were carried out using a Bush analog computer and a cadre of over 100 women calculating by hand. The ideas of Mauchly and Eckert were embodied in the first digital computer, the electronic numerical integrator and calculator (ENIAC). The ENIAC took over two and a half years to build, contained 18,000 vacuum tubes, and could complete 5,000 additions per second using electronic impulses.

After the war the original Aberdeen group split up and Mauchly and Eckert decided to produce commercial computers. The firm of Eckert–Mauchly Computer Corporation later became a division of Remington Rand, which in 1951 introduced the UNIVAC I, first installed in the Bureau of the Census. UNIVAC I improved upon the original ENIAC design by incorporating the concept of the stored program and conditional transfer originated by John von Neumann, a renowned Hungarian-born mathematician, and others who worked with him at Aberdeen during World War II. The modern computer industry was created from these beginnings, its patralineage frequently traceable to government support through the numerous projects conducted under the auspices of the Aberdeen Proving Ground.

CONCLUSION

The three cases presented here, spanning over a 150 years of our national economic experience, suggest a powerful underlying reason for the growth and prosperity of the American manufacturing sector. In none of the examples referred to did the government seek to regulate or restrain the firms or individuals who sought to apply the benefits of government-sponsored innovation. In the absence of such restraint, normal market incentives ensured the fullest economic development of each new product or process. Today, as in the past, the national government possesses enormous, legal, and economic power—power that is indispensable in the creation of new goods and productive techniques. The disposition of that power has had and will continue to have a powerful influence on the rate and direction of inventive and innovative activity. With the government acting in a facilitative and supportive fashion, as it has throughout most of our history, we would be better positioned to face the competitive challenge of the global economy.

23

Competitiveness and the Antieconomics of Decline

DONALD N. McCLOSKEY

Start with a riddle, by guessing the time and place:

A nation speaking the language of Shakespeare wins a world war and takes command of the balance of power. It builds the largest economic machine in history, and is acclaimed on all sides as having the most energetic businesspeople, the most ingenious engineers, the smartest scientists, and the wisest politicians. Then everything goes to hell. An upstart challenges its economy, beating it at its own game. The former paragon, a decade or two after the hosannas, comes to be scorned, at any rate at home, as having the laziest businesspeople, the stupidest engineers, the dullest scientists, and the most foolish politicians. It becomes in the opinion of the world (or at any rate in the opinion of homefront journalists and politicians) a New Spain or a New Holland, a byword for a failed empire.

Time's up.

ANSWERS, WRONG AND RIGHT

If you guessed "America, 1916–1992," give yourself half credit, fifty points. Sorry, that's not passing. True, the story fits American history from The War to End All Wars down to the Noriega Trial, but it also fits more. If you guessed "Britain 1815–1956," you again earn fifty points, and ten points extra for recognizing that there is a world outside the United States. The story fits British history, all right, from Waterloo to Suez, but the answer warrants only a pass, D −. The better, cum laude answer is "Both, down to details of the words people used at the time to describe what was happening." British opinion leaders in the 1890s and 1900s read books with titles like *Made in Germany* (1886) or *The American Invasion* (1902). Where have you seen those? You've seen them in airport bookstores from

Kennedy to Honolulu, with America in place of Britain as the patsy. The book of 1896 might as well be reprinted 100 years later with "Japan" in place of "Germany": "Make observations in your surroundings. . . . You will find that the material of some of your own clothes was probably woven in Germany . . . the toys and the dolls and the fairy books" and the piano, the drain-pipes, and on and on, down to a souvenir mug inscribed "A Present from Margate, Kent." Make observations in your garage, your living room, you den; you will see Toyota, Sony, and Yamaha all around.

The answer, however, is too clever by half. Give yourself a B−, and try to be a little wiser next time. Sure, warning that "The End is Near" gives one a reputation for prudence, besides selling newspapers. The story is made better by a supposedly horrible example, such as Britain, the only European country many Americans think they know, and therefore think they know how to improve. The British and American stories certainly are told in parallel. The British analogy haunts American intellectuals. On the upside each country in succession became the world's banker. On the downside both fought a nasty colonial war against farmers (Boer and Vietnamese). Both in the end became debtor nations, with long-lasting deficits in visible trade.

The stories, however, are wrong, both of them. That is the correct answer for an A+ and an invitation to graduate school. The stories are routinely applied to Britain and now America, but they are mistaken. As much as American intellectuals delight in telling them around the fern bar, urging us to buckle up our football pads for *The Zero-Sum Solution: Building a World Class Economy* (L. Thurow, 1985) or to finally get down to *The Work of Nations: Preparing Ourselves for 21st-Century Capitalism* (R. B. Reich, 1991), the story is wrong about America. It was just as wrong about Britain a century ago. The story of *The Rise and Fall of the Great Powers* (P. Kennedy, 1987) is a fairytale. The correct story is that both countries were and are economic successes.

THE MYTH OF DECLINE

Here are the data. Angus Maddison is a Scot living in France and working in Holland. He is a bear of man fluent in seven languages and in statistical thinking, and is the leading authority on the history of world trade and income. In 1989 he published a little-noticed pamphlet entitled *The World Economy in the 20th Century,* under the auspices of the Organization for Economic Cooperation and Development (OECD), the research institute in Paris for the industrial countries. Using the best statistics on income available Maddison came up with the following surprising facts:

1. Americans are still richer than anyone else, after a decade of "failure." In 1987 Americans earned $13,550 per head (in 1980 prices), about 40 percent higher than, say, the Japanese or the (West) Ger-

mans. If you do not believe it have a look at a typical house in a suburb of Tokyo.

2. Britain is still rich by international standards. After a century of "failures" the average Briton earns a trifle less than the average Swede and a trifle more than the average Belgian. If you do not believe it, stay at a Belgian hotel. The British average, however, is over three times that of Mexico and fourteen times that of India. If you do not believe it, step outside your hotel in Calcutta.

America, therefore, has not "failed," and neither has Britain.

The American story as it is told in the lecture rooms repeats the British story, eerily, but it is a matter of false rhetoric from the start. British observers in the early nineteenth century, like Americans in the Jazz Age, were startled at the ease with which the country had taken industrial leadership. Britain was the first, but a few of its intellectuals were nervously aware of the strangeness of a small island running the world. In 1840, early in British success, J. D. Hume warned a select committee of Parliament that tariffs on imports of wheat would encourage other countries to move away from agriculture and toward industry themselves, breaking Britain's monopoly of world manufacturing:

> [W]e place ourselves at the risk of being surpassed by the manufactures of other countries; and . . . I can hardly doubt that [when that day arrives] the prosperity of this country will recede faster than it has gone forward.

Nonsense. It is the "competitiveness" rhetoric, and it has always been nonsense, in the 1840s or the 1990s. Britain was made better off by the industrialization of the rest of the world, in the same way that you would be made better off by moving to a neighborhood of more skilled and healthy people. British growth continued from 1840 to the present, making Britons richer and richer. Likewise, Americans are made better off when Japan "defeats us" at carmaking because we then go do something we are comparatively good at—banking, say, or growing soybeans—and let the Japanese do the consumer electronics. Richer and richer. According to Maddison, Britain is about three-and-a-half times richer than it was a century ago; America about five times richer.

It is true that Britain and America have grown slower than some other countries, probably because Britain and American started richer. The story of industrial growth in the past century has been a story of convergence to British and American standards of excellence. Germans in 1900 earned about half of what Britons earned; now they are about the same. It is not a "race" that Britain lost. The falling British share of world markets was no index of "failure," any more than a father would view his falling share of the poundage in the house relative to his growing children as a "failure." It was an index of maturity. This was also true for America. It is good, not bad, that other nations are achieving American standards of competence in running supermarkets and making food processing equipment. Three cheers for foreign "competition."

The story of "failure" has consequences itself, which is why it needs to be challenged. It confuses prestige of a sporting character, being tops in what the British call the "league tables of economic growth," with significant differences in wealth. More ominously, it speaks of free exchange in metaphors of war. In 1902, at the height of xenophobic hysteria in Britain about "competitiveness," Edwin Cannan declared

> [T]he first business of the teacher of economic theory is to tear to pieces and trample upon [hold on there, Edwin] the misleading military metaphors which have been applied by sociologists to the peaceful exchange of commodities. We hear much . . . in these days of "England's commercial supremacy," and of other nations challenging it, and how it is our duty to "repel the attack," and so on. The economist asks "what is commercial supremacy?" and there is no answer."

We hear much these days of America's commercial supremacy and how it is our duty to repel the attack. In 1884–1914 such talk led to a world war. In our times we should perhaps cool it, recognizing for instance that most jobs lost in Massachusetts are lost to Texas and California, not to Japan and Korea; or that richer neighbors will pay more to us for our goods and services.

David Landes, a professor of history and of economics at Harvard, brought the mistaken story of Britain's decline to academic respectability. In his classic book, *The Unbound Prometheus: Technological Change and Industrial Development in Western Europe from 1750 to the Present* (1969), Landes summarized a century of journalistic and historical weeping for lost supremacy and lost empire. He uses throughout a metaphor of leadership in a "race," speaking in chapter titles of "Closing the Gap" and "Short Breath and Second Wind," with a military version in "Some Reasons Why," taken from a poem about a cavalry charge.

The main question, according to Landes, is, "Why did industrial leadership pass in the closing decades of the nineteenth century from Britain to Germany?" Briefly, his answer is:

> Thus the Britain of the late nineteenth century basked complacently in the sunset of economic hegemony. . . . [N]ow it was the turn of the third generation, the children of affluence, tired of the tedium of trade and flushed with the bucolic aspirations of the country gentleman. . . . [T]hey worked at play and played at work.

That is fine writing, but it merely restates the nonsense about competition. It is nonsense on two grounds, political and economic. The European story is in fact commonly told by diplomats and their historians in terms of footraces and cavalry charges among ironmasters and insurance brokers, and the sunset of economic hegemony. The balance of political power in

Europe since Peter the Great is supposed to have depended on industrial leadership. Waterloo and the Somme are supposed to have been decided on the assembly line and trading floor. The supposed link between the lead in war and the lead in the economy became a commonplace of political talk before World War I and has never since left the historical literature. To think otherwise, says Landes, is "naïve."

The link between the economy and politics, it needs to be said, is nonsense. After all, a large enough alliance of straggling, winded followers could have fielded more divisions in 1914. The case of Soviet Russia in 1941 or North Vietnam in 1968 suggest that military power does not necessarily follow from economic power. In 1861–1865 the Union sacrificed more men than the entire United States did in any other war to put down a rebellion by a less populous section than it outproduced at the beginning by 30:1 in firearms, 24:1 in locomotives, and 13:1 in pig iron. In World War I the shovel and barbed wire, hardly the most advanced fruits of industry, locked the Western Front. Strategic bombing, using the most advanced techniques and the most elaborate factories, failed in World War II, failed in Korea, and was therefore tried again with great fanfare, to fail again, in Vietnam. It worked finally against a trivial military power in Iraq. Or did it? The equation of military power with economic power is good newspaper copy, but it is poor history.

BEYOND THE LOMBARDI METAPHOR

The economic nonsense in the metaphor of leadership is that it assumes silently that first place among the many nations is vastly to be preferred to second or twelfth. Leadership is number-oneship. In the motto of the great football coach, Vince Lombardi, "Winning isn't the most important thing; it's the only thing."

No. The metaphors of disease, defeat, and decline are too fixated on Number One to be right for an economic tale. The Lombardi motto governs narrowly defined games. Only one team wins the Super Bowl. The fixation on Number One, though, forgets that in economic affairs being Number Two, or even Number Twelve, is very good indeed. The prize for second in the race of economic growth was not poverty. The prize was great enrichment. In other words, since 1870, Britain has grown pretty damned well, from a high base.

By contrast, the diseases of which the pessimists speak so colorfully are romantically fatal; the sporting or military defeats are horribly total; the declines from former greatness irrevocably huge. A historian can tell the recent story of the first industrial nation as a failure, and be right by comparison with a few countries and a few decades. The historian would sell plenty of books to Americans in the last years of the twentieth century because some Americans worry about "loss of leadership." The historian, however, would be writing nonsense.

On a wider, longer view the story of failure in a race is strikingly inapt. Before the British the Dutch were the "failure." The Dutch Republic has been "declining" practically since its birth. With what result? Disaster? Poverty? A "collapse" of the economy? No. The Netherlands has ended up small and weak, a tiny linguistic island in a corner of Europe, stripped of its empire, no longer a strutting power in world politics—yet fabulously rich, with among the highest income in the world (now as in the eighteenth century), a domestic product per head quadrupling since 1900, astoundingly successful by any standard but Lombardi's.

The better story is one of normal growth, in which maturity is reached earlier by Britain and America than by Japan and Germany. The British failures of the late nineteenth century were small by international standards, even in industries such as steel and chemicals in which Britain is supposed to have done especially badly. Everyone concedes that in shipbuilding, insurance, bicycles, and retailing Britain did well from 1870 to 1914. Whether it "did well or not," however, its growth did not depend importantly on keeping right up with Number One. Britain in 1890 could have been expected to grow slower than the new industrial nations. The British part of the world got there first, and was therefore overtaken in rate of growth by others for a time. Naturally, someone who already passed the finish line is going to be moving slower than someone who is still running. Belgium was another early industrial country and had a similar experience of relative "decline" that is seldom noted.

On the whole, with minor variations accounted for by minor national differences in attention to detail, the rich nations converge. Resources are a trivial element in modern economies. Technology, on the other hand, has become increasingly international. If people are left to adopt the most profitable technology, then they end up with about the same income, whether they live in Hong Kong or Dresden.

The main British story since the late nineteenth century is what Americans can expect in the century to come. British income has tripled while others achieved British standards of living. A 228 percent increase of production between 1900 and 1987 is more important than an 8 percent "failure" in the end to imitate German habits of attention to duty. Looked at from Ethiopia or even from Argentina, Britain is one of the developed nations. The tragedy of the past century is not the relatively minor jostling among the leaders in the lead pack of industrial nations. It is the appalling distance between the leaders at the front and the followers at the rear.

If one must use the image of the race course, then the whole field, followers as well as leaders, advance notably—usually by factors of 3 or more since 1900 in real output per head. The main story is this general advance. The tripling and more of income per head relieved much misery and has given life affording scope to billions of people otherwise submerged: Think of your great grandparents.

In other words, the trouble with this pessimistic choice of story in the literature of British and American failures is that it describes this happy outcome of growth as a tragedy. Such talk is at best tasteless in a world of real tragedies—Argentina, once rich, is now subsidizing much and producing little; or India, trapped in poverty after much expert economic advice. At worst the pessimism is immoral, a nasty self-involvement, a line of nationalist guff accompanied by a military band playing "Land of Hope and Glory" or "The Marine Hymn." The economists and historians appear to have mixed up the question of why Britain's income per head is now six times that of the Philippines and thirteen times that of India—many hundreds of percentage points of difference that powerful forces in sociology, politics, and culture must of course contribute to explaining—with the more delicate and much less important questions of why British income in 1987 was 3 percent less than the French or 5 percent more than the Belgian.

CONCLUSION

Telling a story of America following Britain into "decline" is dangerous nonsense. It is nonsense because it is merely a relative decline, caused by the wholly desirable enrichment of the rest of the world. It is dangerous because it leads us to blame foreigners for our real failings, in high school education, say, or in the maintenance of bridges.

So cheer up. We are not going the way of Britain, if that means what the pessimists mean. In the more accurate and optimistic story it means continuing to succeed economically. If success means cricket on Sundays and drinkable beer, then hip, hip, hurray.

24

Does Science Shape Economic Progress—Or Is It the Other Way Around?

NATHAN ROSENBERG

It is widely believed that an industrial civilization, such as late twentieth-century America, depends upon scientific research for its continued economic success. Although there is no reason to doubt the long-term validity of this belief, it is often associated with a striking unawareness of the considerable extent to which new technologies can emerge *without* the assistance of recently acquired scientific knowledge. Indeed, to think of the lines of causation as running exclusively from new scientific findings to improved economic performance is highly selective and incomplete, and can lead to incorrect conclusions for the making of public policy.

STEAM ENGINES AND SILKWORMS

The basic deficiency with the view that scientific advances are a cause, and economic improvement a consequence, is that it never even poses the question of what *brings about* scientific progress in the first place. Friedrich Engels made the point forcefully, perhaps even *too* forcefully, in a letter written in 1895:

> If, as you say, technique largely depends on the state of science, science depends far more still on the *state* and the requirements of technique. If society has a technical need, that helps science forward more than ten universities.*

One does not need to believe, as Engels did, that political economy wholly determines scientific advance, in order to believe that the sphere of pro-

*Engels to H. Starkenburg, January 23, 1895; Engels' emphasis.

174

duction has long shaped the direction of scientific activity in powerful ways. The growth of science as a profession, as has taken place in the past century or so in universities and research institutes, has certainly meant that research problems are regularly formulated for their intrinsic interest, or to satisfy the persistent curiosity of an inquiring scientist, or for their potential importance in clearing up anomalous observations or apparent inconsistencies. The scientific enterprise obviously has a forceful internal dynamic of its own. Nevertheless, even here, the potential utility of scientific findings in certain spheres is an important argument in determining the allocation of tax monies in support of research in particular fields. Nor is this a recent development in history; publicists for the cause of science since the time of Francis Bacon, almost 400 years ago, justified these activities on the grounds of an eventual flow of material goods.

All of which simply means that events in the economic sphere have played a major role in shaping the *agenda,* and therefore the *eventual* findings, of pure science. In this sense, the lines of causation and influence run from economics to science as well as from science to economics. Indeed, some of the most fundamental scientific achievements of the nineteenth century need to be understood in precisely these terms.

Sadi Carnot's creation of the modern science of thermodynamics, in the 1820s, was the outgrowth of an attempt to explain what it was that determined the efficiency of steam engines. This is clear in the title that he chose for his immensely influential book, published in 1824, *Reflections on the Motive Power of Fire and on the Appropriate Machinery for Developing that Power*—which was published, it is important to note, about fifty years *after* James Watt's seminal work in developing the steam engine.

The Englishman James Prescott Joule's discovery of the law of the conservation of energy grew out of his interest in alternative energy sources for the generation of power in his father's brewery. Similarly, the great Pasteur formulated the modern science of bacteriology in the 1860s. This formulation emerged directly out of his research on fermentation processes in accounting for the deterioration of French wine, and the cause of the silkworm disease that threatened the very survival of the French silk industry.

A SECOND LIFE

Economic issues have not only played a crucial role in the formulation of the research problems that have shaped the problems of science, but they have also played a decisive role in shaping the eventual impact of scientific breakthroughs, once they have occurred. The practical implementation of a given technological advance typically depends upon the willingness to commit large amounts of resources to highly risky activities—as well as a willingness to wait for long periods of time before realizing any sort of financial return. These investments are the *D* of R&D (Research and

Development). In the United States in recent years these development expenditures have accounted for more than two-thirds of all R&D. It is not nearly as widely understood as it ought to be that R&D expenditures are primarily upon D rather than R. Basic and applied research have constituted less than one-third.

Consider the excitement over recent breakthroughs in superconductivity. As a purely scientific breakthrough, the excitement is well justified, but it will probably be a matter of decades before this is actually translated into better computers, magnetically levitated trains, the transmission of electricity without loss, or the storage of electricity. It will be extraordinarily difficult, time-consuming, and expensive to design new products that exploit new knowledge of high-temperature superconductors, and then design and market the technology that can produce these new products.

The state of scientific knowledge about superconductivity today is perhaps comparable to that concerning electricity when Faraday discovered the phenomenon of electromagnetic induction in 1831. With the important exception of the telegraph, Faraday's discovery did not give rise to any major technological innovations for several decades. It took more than forty years before Maxwell's formulation brought electrical phenomena to a level of understanding that was not only systematic and mathematical, but also had specifically testable implications; it was another few decades beyond that before the pioneering work of Marconi and the use of radio waves for long-distance communication.

In some cases, new scientific understanding has been so remote from a capability for exploiting it in an economically meaningful way that an entirely new discipline has to be created to bring this about. Such was the case toward the end of the nineteenth century in chemistry, and the end result was the development of the new discipline of chemical engineering. Perkins' accidental synthesis in 1856 of mauveine, the first of the synthetic aniline dyes, rapidly gave rise to a new synthetic dyestuffs industry and exercised a powerful impact on research in organic chemistry. Discovering or synthesizing a new material under laboratory conditions, however, was only the very first, tentative step toward the possession of a marketable final product. Devising process technologies for producing new chemical products on a commercial basis required years of work and large financial commitments by profit-driven establishments such as the Du Pont labs.

The general point is that any innovation, once made, and regardless of the extent to which science contributed to its beginnings, takes on a second life in which the eventual outcome is shaped by commercial and, more broadly, economic forces. Understanding the scientific and technical basis for wireless communication, which Marconi did, was a very different matter from understanding the possibilities for a vast new entertainment broadcasting industry that had the capability for reaching into every home and automobile in the country. Marconi failed completely to envisage these possibilities. They were much better understood by David Sarnoff, who eventually became president of RCA. Sarnoff had no professional training of any kind, but he had an immensely greater intuitive sense for the long-run commercial possibilities of the new radio technology. Economic impact or social significance are not things that can be extrapolated out of a piece of hardware.

The impact of new technologies will depend on what is subsequently designed and constructed with them. New technologies represent unrealized potentials that may take a very large number of eventual shapes. What shapes they actually take will depend on a wide range of social priorities and values as well as on the way the demand for particular goods and services changes in response to rising incomes or declining prices. Gunpowder was discovered by the Chinese, who used it for festive and celebratory purposes. When Europeans acquired this technology they quickly turned it to very different ends, and dramatically altered world history.

DOWNSTREAM DEVELOPMENT

A special feature of inventions is that they commonly arise in the attempt to solve some very specific, narrowly defined problem. Once found, however, the solution often turns out to have important applications in other, totally unintended contexts.

The steam engine, for example, was invented in the eighteenth century specifically as a device to pump water out of flooded mines. It owed nothing, as we have already seen, to an understanding of the laws of thermodynamics. For a long time, it was regarded exclusively as a pump. A succession of improvements later rendered it a feasible source of power for textile factories, iron mills, and industrial establishments. In the course of the early nineteenth century the steam engine became a generalizable source of power and had major applications in transportation. Later in the nineteenth century the steam engine was used to produce a new and even more generalizable source of power—electricity—which, in turn, satisfied innumerable final uses to which steam power itself was not efficiently applicable. Finally, the steam turbine displaced the steam engine in the generation of electric power, and the special features of electricity—its ease of transmission over long distances, the capacity for making power available in "fractionalized" units, and the far greater flexibility of electrically powered equipment—spelled the eventual demise of the steam engine.

The history of the steam engine was thus shaped by forces that could hardly have been foreseen by inventors who were working on ways of removing water from increasingly flooded coal mines. Its subsequent history was shaped by unanticipated applications to industry and transportation and eventually by the systematic exploitation of new technologies, such as electricity, that were undreamed of when the steam engine itself was invented. Nevertheless, the steam engine played an important role in bringing about the commercial exploitation of this new science-based technology. It also played a vital role in one of the greatest of all scientific discoveries: Carnot's formulation of the second law of thermodynamics.

A related point is that major innovations, such as the steam engine, have the effect of inducing further innovation and complementary investments over a wide frontier. Indeed, the ability to induce further innovations and investments is a reasonably good definition of what constitutes a truly major innovation. It is a useful way of distinguishing between technological advances that are merely invested with great novelty from advances that have the potential for a major economic impact.

It is evident, then, that ongoing activities within the economic realm are highly influential in shaping the advance of science. There is, however, another dimension to these relations that needs to be highlighted, although it is implicit in what has already been said. That is, even though there are good reasons to expect that economic benefits will continue to flow from scientific research, it has become less likely that such benefits will provide competitive economic advantages to the country conducting such research. Whether or not it will do so will depend to an increasing degree on the country's capabilities for downstream development.

In this regard, compare the recent experiences of Great Britain and Japan. Great Britain's lackluster economic performance in the years since World War II demonstrates the insufficiency of high-quality basic science when it is not closely associated with complementary commercial and engineering skills, and when the economic environment fails to offer sufficiently high rewards to technological innovators or to the adopters of newly available technologies. For many years after World War II, Great Britain received more Nobel Prizes in science, on a per capita basis, than did America, yet failed to convert this scientific edge into economic advantage. Japan, on the other hand, has received very few Nobel Prizes in science. Indeed, Japan has received only a very small fraction of the number of Nobel Prizes received by a single British institution—the Cavendish Lab. Nevertheless, it would be difficult to make the argument that this relatively unimpressive performance at the forefront of scientific research has been a major handicap to the Japanese economy. Rather, the Japanese experience forcefully demonstrates the great possibilities for economic growth based upon the systematic transfer and exploitation of foreign technologies.

CONCLUSION

It seems obvious from these examples that, given the appropriate engineering, managerial, and organizational skills, a high degree of economic performance can be attained by drawing upon more advanced technologies that are available abroad and that are the product of scientific research conducted abroad. To put it the other way around, it is easy to exaggerate the purely economic significance of a first-rate domestic scientific research capability when it does not exist together with the complementary market savvy. To be more precise, scientific capabilities are certainly not a sufficient condition for economic competitiveness and growth. They are becoming more frequently a *necessary* condition, if only because earlier access to new research findings can be translated into lead time over one's competitors that may be highly advantageous in commercial terms.

Appendix:
Chronology of Important Dates

301 Price and wage ceilings imposed by Roman Emperor Diocletian, wreaking havoc on an already deteriorating Imperial economy.

1623 Maximum commodity prices issued by governor of Virginia.

1700 to 1754 British military expenditures averaged about 10 shillings per person per year.

1700 to 1775 Trade deficit averaged close to 4 percent of colonial income.

1717 Isaac Newton, British master of the Mint, set too high a price on silver, transferring Britain to de facto gold standard on which she would remain for more than two centuries.

1730 Pennsylvania Assembly refused to spend any money to stop privateers from interfering with the colony's trade.

1756 to 1763 When Britain fought the French and their allies on the North American continent, British military expenditures more than doubled.

1756 to 1776 To cover extra costs of colonial defense, British taxpayers faced the highest levies in the Western world—about £1 per person per year, or about 6 percent of the island nation's net national product.

1770s James Watt developed steam engine.
Mechanical inventions in cotton spinning made it impossible for a hand spinner to earn a living wage, but there were growing demands and good wages available for British workers willing to work the mechanical spinners.

1771 American trade deficit was over 9 percent of total colonial income, compared to less than 3.5 percent of U.S. income in 1986, the worst year of the 1980s.

1773   American Revolution sharply spurred by colonial reactions to the granting of a monopoly in tea to the English East India Company.

1776   Accumulated debt of the colonies to Britain alone amounted to $40 million
Declaration of Independence complained that King George III was hindering the peopling of the colonies from abroad. The population of the thirteen states was about 4 million.
Settlers pushed into American frontier lands, raising issue of whether land should be sold to generate revenue, or given away. Alexander Hamilton was major exponent sales for revenue; Thomas Jefferson was the leading advocate of free-land policy.

1776 to 1781   Faced with hyperinflation, the Continental Congress set maximum prices, leading to black market in scarce goods and diversion of supplies to regions where prices were free.

1776 to 1913   United States relied on tariff as major revenue source.

1776 to 1781   Revolutionary War. Total direct cost in annual contemporary gross national product (TDC/GNP): 104 percent. Only 10–15 percent of cost was covered by current taxation.

1787   At Constitutional Convention, James Madison recognized link between liberal immigration policy and economic growth.

1789   Constitution of the United States expressly forbade exclusive monopoly privileges based on state patents.
During his diplomatic assignments in France, Thomas Jefferson picked up the notion of interchangeable parts in factory production.

1790   British workers lured into cities by better jobs.

1790s   President Washington established two federal armories to store, repair, and produce limited quantities of arms to ensure that there would be a secure supply of ordinance in time of need.

1791   Treasury Secretary Alexander Hamilton urged that foreign capital was good, not bad.
The First Bank, proposed by Hamilton on the model of the Bank of England, was chartered for twenty years by Congress.

1792   "Buttonwood Tree" agreement established fixed commissions and trading priorities for New York stock trades.

1793   Thomas Jefferson was instrumental in securing an arms contract for Eli Whitney; Whitney was obligated to produce 10,000 stand of arms in two years, made of fully interchangeable components.

1800s   During this century there were only twenty years when the United States did not run a foreign trade deficit. The deficits exactly mirrored booms and slumps in transportation investment.

Prices rose and fell, ultimately cancelling each other out: the price level in 1900 was roughly what it had been in 1800.

With massive silver discoveries, real price of silver fell relative to official price.

1800s, early Steam engine became generalizable source of power, had major applications in transportation.

1800s–1850s Irish immigrants in New York and Boston seen as unassimilable degenerates.

1810 Fulton schemed to win exclusive control over all steam navigation throughout the entire trans-Appalachian West; pulled political strings.

1810–1830s Freight traffic out of the vast midwestern interior through New Orleans accounted for over 80 percent of all outward shipments.

1810–1860 Average number of yearly round trips per vessel quadrupled. Total outbound freight tonnage from the country's interior growth grew from 65,000 tons to 7,690,000 tons. The population of the trans-Appalachian west grew from 1 million to 13 million, from 15 to 41 percent of the total population.

1811 First Bank's charter expired.

Territory of Orleans legislative assembly, granted exclusive steam navigation rights to Fulton, who launched 371-ton *New Orleans,* at Pittsburgh. Fulton also petitioned the Russian czar for steamboating monopoly rights between St. Petersburg and Kronstadt.

1812 Income tax first proposed during war with England; war ended before Congress could vote on proposal.

Fulton signed an agreement with an Englishman to launch steamboating on the Ganges River in India.

1812–1815 War of 1812.

Without the First Bank, federal finances became chaotic.

Chief of Ordinance reorganized federal armories, moving the system of production away from organized gunsmithing to something reflecting modern manufacturing practice, characterized by the use of special-purpose machinery and division of labor.

1815 Steamboat of mechanic and inventor Henry Shreve arrived in New Orleans; his vessel was seized and then released on bail, but other ships were impounded.

1816 Congress chartered Second Bank of the United States.

Louisiana State Supreme Court ruled steamboat monopoly was unconstitutional, but local politics remained in Fulton's favor.

1817 Anticipating a fairly sure loss at U.S. Supreme Court, Fulton's heirs abandoned monopoly fight; open competition on the western rivers was assured.

Francis Cabot Lowell developed working version of British mechanical loom.

1817–1820   Freight rates (in 1820 constant dollars) per 100 pounds shipped upstream from New Orleans to Louisville fell from $3.12 to $2.00.

1819   Fifty-nine steamboats were plying the Western trunk rivers.

1820   Crude birth rate in United States was 55.2 per thousand.
While inequality rose in Britain after this time, the share of saving in national product failed to rise at all.

1820–1860   Freight rates (in 1820 constant dollars) per 100 pounds shipped upstream from New Orleans to Louisville fell from $2.00 to $0.28.

1821   Boston investors (Merrimack Manufacturing Company) built factory town at Lowell, Massachusetts.
Crude birth rate in England: 40.8 per thousand.

February 1822   Merrimack investors launched "Lowell Experiment," building factory dormitories and offering New England farm girls clean working conditions, supervised living, and pay that was more than these women could earn in any other occupations open to women at the time. Women rushed to Lowell.

1823   Second Bank became central reserve, stabilizing the monetary and banking systems, while continuing to compete with state banks for commercial business.

1824   Sadi Carnot published *Reflections on the Motive Power of Fire and on the Appropriate Machinery for Developing that Power;* created modern science of thermodynamics.

March 1, 1826   City of Lowell, population 2,500, was incorporated.

1830s   At a time when young women were not allowed to attend college, the Merrimack company arranged for Harvard University professors to come to Lowell to address the factory women. These women even edited a literary magazine, the *Lowell Offering,* which became famous throughout the United States and Europe.
By now, two workers on a power loom in a factory could produce, in one day, twenty times what a hand-loom weaver could produce at home. Displaced weavers (Luddites) burned factories and smashed machines.
Jackson administration backed greater democracy in banking politics and freer competition in banking enterprise.
National armories became state-of-the-art facilities which had achieved practical interchangeable manufacture, the forerunner of modern mass production.

1830s to 1860s   U.S. economy grew rapidly.

1830s to 1890s   The savings rate rose markedly making it possible to finance the enormous investment requirements of industrialization and urbanization.

1832   Congress voted to renew the Bank's charter.

1834 to 1844   Sixty percent of investment in canals came from foreign sources.

1835 National debt retired; Federal land giveaways gained favor.

1836 Federal bank expired.

1836 Ten years after incorporation, the population of Lowell had swelled by more than 500 percent, to nearly 18,000.

1830s, late So-called free-banking laws emerged.

1837 to 1914 United States had no central bank.

1837 to 1980s No interstate branch banking in the United States.

1839 Market for American securities slumped in London; work on capital projects was checked throughout eastern United States, and stopped entirely in much of West.

1840 to 1920 Immigrants, instead of being an underpaid, exploited group, generally held an economic position that compared favorably to that of native-born Americans.

1840s European famine during failure of Irish potato crop.
Faced with competition from other mills, Lowell investors instituted incentives to speed production. Work that had been difficult at the earlier wages now became intolerable to many women. Some launched strikes and tried to persuade the Massachusetts legislature to institute mandatory ten-hour day.

1848 After potato famine, thousands of Irish families migrated to Lowell.

1849 Western river steamboats logged 3.3 billion freight-ton miles and 1.1 billion passenger miles—1 billion ton-miles more than all nation's railroads were hauling ten years later.

1850s Most states had adopted free-banking laws; nationwide system based on correspondent relationships between banks in different states began to emerge.

1856 Perkins' accidental synthesis of mauveine exercised a powerful impact upon research in organic chemistry.

1860 Adderly suggested that financial assessment imposed on the British taxpayer by the colonies should no longer be countenanced.

1860s Louis Pasteur formulated modern science of bacteriology.

1861 to 1865 During Civil War, Congress adopted income tax, accounting for nearly one-third of all internal revenue; nearly 60 percent of revenue came from New York, Massachusetts, and Pennsylvania.
TDC/GNP: 74 percent and 123 percent for Union/Confederacy, respectively. Union raised tariffs, levied excises and, for first time, taxed incomes.

1862 Government of Gambia and British Colonial Office disputed over expenses incurred in connection with the war against King Badiboo.

Homestead Act made land giveaway explicit U.S. policy objective.

Morrill Act granted land endowments to state colleges offering courses in agriculture and mechanic arts.

1863 Federal government adopted deregulated free-banking system.

1865 to 1893 Foreigners provided more than one-third of the financing for railroads, buying up 50 percent of bonds, 25 percent of stocks issued by railroad companies.

1865 to 1900 The rate at which Southern blacks accumulated wealth—in real estate and other forms—exceeded the rate at which Southern whites accumulated wealth.

1870 to 1960 Real income per person in America as a whole rose fivefold; the incomes of black people not only matched this dizzy growth, but surpassed it.

1870s Holland, Denmark, Norway, and Sweden suspended silver coinage and turned to gold, an example quickly emulated by France, Belgium, Switzerland, Italy, and Greece.

Reassertion of railroad investment; foreign funds transformed United States from capital-scarce to capital-rich economy.

1870s to 1880s Observers in Britain complained of profit squeeze, and historians coined the name "Great Depression" to characterize the state of the British economy.

1800s, late State legislatures began passing "employer liability laws" that either eradicated employers' standard defenses or made it easier for workers to obtain accident compensation. Prior to this time, liability for workplace accidents had been based on commonlaw standards of negligence.

1800s, late, through 1930s Ring spindle eliminated need for most skilled or strong loom workers, allowing southern mills to take over the entire textile market. With the movement of the textile industry from Lowell to the South, the northern mills were used to make carpets, shoes, sweaters, and various other industrial products.

1871 to 1873 Germany used indemnity from Franco–Prussian War to purchase gold and establish gold-based currency unit.

1872 War-spawned income tax abolished, but high tariffs remained.

1873 to 1893 U.S. prices fell by about 1 percent per annum.

1873 to 1916 Under various modifications of Homestead Act, 285 million Federal acres transferred to individuals; Federal restrictions and development costs amounted to as much as 80 percent of the total value of lands.

1875 Britain cajoled Straights Settlements into accepting fiscal responsibility for all of the costs of Perak War.

1875 to 1900 Expansion of U.S. exports and foreign investment created interest groups that favored freer trade and revival of income tax as alternative to import tariffs.

1877 U.S. Supreme Court held (in *Munn* v. *Illinois*) that any private property "clothed in the public interest" was lawfully subject to government regulation.

1879 to 1910 Veterans' pension payments and military expenditures increased from one-third to two-thirds of federal budget.

1879 United States restored gold convertibility; Russia and Japan also embraced gold standard.
Zulu War of 1879 broke out in Natal Colony; only one-fourth of the £1 million cost of the British defense effort was ever squeezed from a reluctant Natal.
American Cigar Makers Union sought to exclude women from factories by law.
Massachusetts passed first enforceable law to restrict the working hours of adult women, setting maximum factory workday at ten hours and maximum work week at sixty hours; this excluded them from much factory work. Labor unions supported the law.

1880s Cyanide process development stimulated directly by real price of gold.
Pension payments tripled, then doubled.
Transcontinental railroadmania; Foreign investment in United States incited nativist sentiment out West. Before this time, trade deficits were rule rather than exception.

1880 to 1890 United States undertook unprecedented peacetime military buildup.

1880s and 1890s Interest payments overseas overwhelmed capital inflows.
Many states followed Massachusetts' lead and passed protective legislation for women workers. Labor unions supported this legislation.

1880 to 1905 Army budget tripled; navy spending went from 6 percent to over 20 percent of U.S. budget

1880s to early 1920s First surge of government growth.

1880 Rutherford Hayes became the first president since Civil War to argue that a larger navy was necessary to protect nation's growing commerce.
Truly international gold standard emerged by this time.

1882 Congressional commission recommended 10–25 percent tariff reduction.

1884 Democratic Party platform opposed the "importation of foreign labor or the admission of servile races."

1887 Anti-Alien Act restricted foreign ownership of land. Hatch Act.
President Cleveland devoted entire State of Union address to attack on high
tariffs.

1887 to 1900 Effects of Anti-Alien Act: Capital investment in mining industry
of the West dried up, while municipal improvements slowed for lack of funds.

1889 Federal land sales, a major source of revenue, began to decline.

1890 About 30 percent of Southern black children ages 5–20 attended some
kind of school.
Because giveaways encouraged premature entry into the frontier, two-thirds of
all homestead claimants before this time were unable to meet Homestead requirements and failed to obtain title to land.

1890s Because of payments to foreign holders of American stock, the economy
continued to run deficits on the current account until the middle of this decade.
Expanding military and veterans' pension programs at the same time that problems were emergent with traditional revenue sources, the Federal government could
either institute a new tax, or it could reduce spending.

1892 American Populist Party agitated for silver coinage to halt decline in
commodity prices; complained country was being crucified on "a cross of gold."
Attacked excise taxes as unfair to farmers and ordinary working folk.

1893 Creation of ICC, passage of Sherman Antitrust Act.

1892 and 1894 American Federation of Labor (AFL) endorsed eight-hour day
for women (which would prevent them from working at most factory jobs) and
tried to outlaw female employment on foot-powered machinery.

1893 to 1913 Trends in gold mining reversed ongoing deflation; U.S. prices
rose by about 1 percent per annum.

1894 First federal deficit since the Civil War; Congress passed income tax despite opposition from northern states, which bore most of tax burden during Civil
War.

1895 Supreme Court ruled that income tax was direct tax and therefore
unconstitutional.
Ilinois Supreme Court struck down an eight-hour day for factory women on
grounds it interfered with right of contract. Labor leaders protested decision.

1895 to 1908 Disproportionate share of Federal spending lavished on northeastern states that opposed income tax, broadening Federal beneficiary base and
pro-tax coalition.

1896 Christian socialist Francis Walker, commissioner general of Immigration
Service and first president of American Economic Association, decried newly arrived "ignorant masses of peasantry."

United States accelerated lending and investment overseas.

In *Plessy* vs. *Ferguson* the Supreme Court established the principle that *de jure* segregated public facilities, including schools, had to be "equal."

1896 to 1915   The volume of net outflows of capital was so large that the United States eradicated its almost three-century history as a net debtor.

1896 to 1972   The current account was always in surplus, giving birth to myth that surpluses are inherently "good" and deficits inherently "bad."

1897   All but 10 percent of accumulated foreign investment was long-term.

1898   AFL asked Congress "to remove all women from government employment . . . and relegate them to the home."

1900   According to the census of this year, 37 percent of Southern blacks between the ages of 15 and 24 were *completely* (not just functionally) illiterate. Average black incomes had risen to a third of white incomes.

Effects of Anti-Alien Act were so clearly seen that an amendment was made that permitted foreign investment in many sectors.

Pennsylvania Supreme Court upheld twelve-hour day law for women; formulated judicial philosophy of women as distinct class in need of protections not needed by men.

1900 to 1930   For every 1,000 men who worked full time in bituminous coal mines in the United States, between three and four were killed in accidents.

1900 to 1940   Relative improvement in the black standard of living slowed.

1900 to 1950   Employment segregation in the South was increasing.

1902   Average British resident contributed fifteen shillings annually toward imperial naval defense; Canadians paid nothing; settlers in Cape Colony, three-and-a-quarter pence; in Natal, five-and-three-quarters pence; New South Wales, eight-and-a-half pence; Victoria and Queensland, one shilling each.

1907   United States had the second-largest battleship fleet in world.

1908   Employer Muller challenged constitutionality of 1903 Oregon ten-hour law for women on grounds that it violated "right to liberty" provided by Fourteenth Amendment; Supreme Court dismissed Muller's challenge, holding that women were "weaker" than men and thus needed special protection. After Muller decision, most states passed legislation to limit women's working hours.

1909   Sixteenth Amendment to Constitution passed in near-unanimous vote.

1910   Estimates of oil losses from fire and evaporation in California ranged from 4 to 8 million barrels, which was 5–11 percent of the state's production.

For every dollar spent per pupil on instruction in Alabama's white schools, only twenty-seven cents were spent per black pupil.

1910 to 1930   Most state legislatures passed workers' compensation laws. These statutes replaced negligence liability with a form of strict liability, under which employers were expected to compensate workers for *all* accidents "arising out of employment." Because employees now had less reason to be careful at work, passage of workers' compensation laws raised the fatal accident rate in the coal industry nearly 20 percent.

1910 to 1960   "Great Migration" of blacks from the South to a small number of major urban centers in the North and the West.

1911   British government urged small Canadian contribution to imperial defense; Canada claimed it was already bearing more than its fair share.

1912   Naval oil reserves created at Elk Hills and Buena Vista Hills in California.

1913   Sixteenth Amendment adopted by states.
United States owed $6.8 billion abroad; then as now, the largest debtor in the world.

1914   Director of the Bureau of Mines estimated losses from excessive drilling in the United States at $50 million, when the value of U.S. production was $214 million.
United States Federal Reserve System founded to provide "stability" in currency markets.

1914 to 1920   Farm prices skyrocketed during World War I and remained high after the war.

1915   Naval oil reserves created at Teapot Dome in Wyoming.

1916   After American entry in World War I inflation was severe; a number of agencies were created to control prices, causing bureaucratic "priorities" escalation.

1916 to 1918   United States in World War I; TDC/GNP: 43 percent. Personal income-tax exemptions reduced from $3,000 to $1,000 for single persons; first-bracket rate increased from 1 percent (1913–1915) to 6 percent (1918), top-bracket rate from 7 to 77 percent.

1918   Cleveland union representing male street car conductors got company to fire all female conductors.

1919   Britain suspended gold convertibility.

1920   Farm prices had more than doubled over the prewar levels.
The proportion in school in the black 5–20 age group had increased to 53 percent and the illiteracy rate was down to 18 percent.
Mineral Leasing Act. To help reduce oil drainage from its lands, the navy authorized small leases for drilling offset wells just inside the borders of Buena Vista and Elk Hills.

July 1920 to July 1921 Farm prices fell 30 percent, never to recover in the 1920s.

1920 to 1970 Federal employment grew 3.5 times as rapidly as U.S. population.

1920s Alice Paul proposed Equal Rights Amendment (ERA), which said that "Equality of rights under the law shall not be denied or abridged by the United States or by any State on account of sex." Unions, women's, and left-wing groups opposed ERA as threat to protective legislation; support for ERA came from Republicans and political right.

Fueled by optimism, farmers expanded holdings and debt.

Magazines such as the *Atlantic Monthly* and *Saturday Evening Post* protested recent waves of immigration from Southern and Eastern Europe.

July 1921 Secretary of the Navy Edwin Denby transferred jurisdiction over naval reserves to the Interior Department to implement new policies for naval oil supplies.

1921 Petroleum executives Doheny and Sinclair allegedly bribed Secretary of the Interior Albert Fall to obtain oil production leases for Naval Oil Reserves at Elk Hills and Teapot Dome.

1922 to 1928 Under pressure from conservationists and small oil companies, Congress reviewed Interior policies.

Mid-1920s New, less-centralized international gold standard reconstructed; this system performed less well than did the nineteenth-century gold standard.

1927 Naval reserve leases cancelled, and the reserves were transferred back to the navy for administration.

1928 On West Texas' Hendrick field, competitive drilling led to construction of storage tanks with a capacity of 11 million barrels at a cost of $3.8 million, while on the neighboring and larger Yates field, where there were larger leases and fewer firms, storage existed for only 783,000 barrels at a much smaller cost of $274,000.

1929 Most households had relatively small amounts of debt.

Private investment was five times as great as the federal budget. Total government expenditures distributed 12 percent of GNP; federal share of GNP was 3 percent.

Stock could be bought on as low as a 10 percent margin: a $10 stock could be bought with a $1 investment and a $9 loan.

With President Hoover's support, Agricultural Marketing Act became law, allotting $500 million for Federal Farm Board price-supports.

October 1929 Stock market lost roughly a third of its value.

1929 to 1930 Decline in U.S. national output was twice as large the typical first year of a recession.

1929 to 1933 Federal Reserve allowed banks to fail and panic to spread without intervening with ready cash to buy up the loans of member banks; banks themselves received much of the blame.

1929 to 1939   Great Depression.
Blacks suffered disproportionately from hardships.
Policymakers failed to harmonize policies adopted in response to Great Depression; severity of deflation and depression led some to suspect gold standard as source of macroeconomic collapse.

1929 to 1987   Government expenditures grew four times as fast as GNP; the federal share alone grew seven times as fast.

1930   Amendment to Mineral Leasing Act of 1920 strongly encouraged, if not mandated, unitized oil production on federal lands; benefits of unitized production incorporated in all federal leases after this time.

1930 to 1990   National opinion polls consistently showed a majority of Americans opposed to increased immigration.

1930s   Efforts to pass legislation compelling oil-field unitization in Oklahoma and Texas.
Influx of Jewish and other refugee scientists and scholars catalyzed American culture and learning.

1931   Farm Board exhausted funds.
President Hoover initiated Senate investigation into securities fraud and abuses.

1931, summer   Epidemic of exchange-rate crises and forced devaluation spread to Europe.

September 1931   Britain abandoned gold standard for final time.

1931, end   Industrial countries began to leave gold standard.

1932   Farm prices had fallen by 50 percent from 1920 levels; boll-weevils, droughts plagued American farmers.

November 1932   Franklin Delano Roosevelt was elected president with a mandate to "do something" about worsening plight of many Americans, including farmers.

1932 to 1934   Twenty-five states passed moratoria on farm foreclosures; U.S. Supreme court acknowledged that these moratoria violated bankers' contract rights, but upheld moratoria on grounds that greater economic good of society justified violation of individual rights to contract. Moratoria made existing loans more costly for creditors; bankers began charging higher interest rates and rationing credit.

1933   Banking Act separated commercial from investment banking; prohibited payment of interest on checking deposits; gave Federal Reserve power to regulate interest on time deposits; established the Federal Deposit Insurance Corporation (FDIC). The goal of the act was to make banking safer by making it less competitive.

Federal government shut down all banks, reopening only those that were financially sound, and insuring the deposits of the reopened banks through the FDIC.

Securities Act; disclosure and antifraud provisions set precedents for future securities regulation.

1933 to 1935 Direct government payments to farmers totaled over $1 billion; indirect costs increased the farm bill to consumers/taxpayers by another $300 million.

1933 to 1937 In part as the result of federal and state agricultural policy, farm failures fell by half, from close to 4 percent to less than 2 percent.

1933 to 1940 Roosevelt increased federal farm aid; taxpayers accounted for one third of some farmers' income; Federal share of farm mortgage debt increased threefold.

1933 to 1990 Price levels increased annually in the United States.

1934 Federal Communications Act created Federal Communications Commission (FCC); set maximum prices for telephone services.

In *Nebbia* v. *New York,* Supreme Court opened floodgates to regulation of *any* private property (discarding the "public interest" limitation of *Munn*).

Securities Exchange Act created Security and Exchange Commission (SEC).

1936 All but four states had hours restrictions that covered women's employment.

1938 Fair Labor Standards Act passed, regulating hours of all workers.

U.S. Navy obtained authorization from Congress to consolidate lands within the reserve and to lease it to a single firm, Standard Oil.

1940 Approximately 9 percent of employed blacks held skilled and white-collar jobs.

1940, summer/fall To assure business of profit from war without much risk, United States changed contracting rules, creating current military procurement establishment.

July 1940 to December 1941 $36 billion made available to the War Department alone—more than the army and navy combined had spent during World War I.

1940 to 1945 American involvement in World War II; TDC/GNP: 188 percent. Expressed in dollars of present-day purchasing power, cumulative war expenditures amounted to more than $2.8 trillion. Personal income-tax exemptions chopped, first-bracket rate went from 4.4 percent (1940) to 23 percent (1944–1945), while top-bracket rate sent from 81.1 percent on income over $5 million to 94 percent on income over $200,000. Payroll withholding instituted.

1940 to 1960s   Prosperity returned to Lowell, Massachusetts, with mobilization for World War II and the growth of the electronics industry after the war.

1940 to 1980   Black–white racial income ratio increased by 12 percent.

1940s   Black–white racial income ratio increased by four percentage points.

1941 to 1946   Longest American experience with controls, during World War II.

1941 to 1991   Income tax accounted for about 60 percent of federal budget.

1941   With onset of World War II, federal spending as share of GNP increased sharply.
Fair Labor Standards Act upheld by Supreme Court.

1942   Members of the Moore School of Electrical Engineering of the University of Pennsylvania, in conjunction with the U.S. Army's Aberdeen Proving Ground, undertook research leading to development of ENIAC (electronic numerical integrator and calculator).

1943   President Roosevelt's famous "Hold-the-Line" order imposed across-the-board controls prohibiting most price increases. Result: black markets, shoddy goods, elimination of lower-priced lines—all of which hit hardest at the poor.
Last great famine in India, in Bengal. Perhaps as many as 3 million died, but food supply dropped by only 5 percent.

June 30, 1944   As of this date, just twenty-six firms enjoyed exactly half the value of all governmentally financed industrial facilities leased to private contractors.

1945 to 1990   Hong Kong's population rose from 700,000 to more than 5 million; during this period its economic development was breathtaking.

1945 to 1991   Tax increases during every year of this period.

1946   When price controls were cancelled, there was a classic example of pent-up inflation—prices were immediately raised.

1947   Compulsory oilfield unitization bill passed in Oklahoma.

1947 to 1950s   West Germany took in unprecedented numbers of immigrants from East Germany and elsewhere; experienced the "German miracle."

1947 to 1990   Cold War costs: 4–11 percent of GNP annually in "peacetime" years.

1949   China's last famine.

1949 to 1991   European taxpayer's per-capita share of NATO defense dropped to 25 percent of average American share.

1949 United States signed North Atlantic Treaty; European taxpayer's per-capita share was only about one-third of the amount that the average American was asked to bear.

1950 By this time, 69 percent of black children were in school, the same as the white attendance rate. The average length of the school year in the South's black schools was 173 days—below the northern average, but fifty-three days more than in 1920.
Crude birth rate in East Asia was 42.5 per thousand.

1950 to 1952 Korean War, a relatively cheap mobilization, was only U.S. war fought on a pay-as-you-go basis. American TDC/GNP: 15 percent. Rapid rise in prices.

1950 to 1980 Federal expenditures grew about half again as fast as GNP.

1950s Pakistani development plans stated that rising inequality might be unpleasant but was essential to finance early industrialization.

1951 Price freeze imposed; taxes raised; Federal Reserve continued restrictive monetary policy.
Remington Rand introduced computer, UNIVAC I, first installed in the Bureau of the Census.

1953 When controls were lifted, there was little postponed inflation.

1954 U.S. Supreme Court reversed separate-but-equal decision in *Brown* v. *Board of Education*.

1956 Justice Department negotiated Consent Decree with AT&T that protected vertically integrated Bell System.

1959 FCC decided Bell System did not have exclusive rights to use of microwave radio.

1960 Forty-one percent of blacks lived in the North, up from 10 percent at the turn of the century. Fully half of postwar increase in the black–white racial income ratio occurred before this time—*before* federal antidiscrimination legislation could have had much effect.

1960 Federal share of government expenditures surpassed the combined shares of state and local governments.
For trades by individuals, $35 million of the $136 million paid in commissions exceeded the competitive rates. Institutions paid total commissions of about $85 million, approximately $55 million of which was above the competitive rates.

1963 Bernie Strassburg became head of Common Carrier Bureau at FCC.

1963 to 1975 Vietnam War. American TDC/GNP: 15 percent.

1964 to 1968   Johnson administration delayed tax increase to finance Vietnam War and resorted to borrowing, pressuring Federal Reserve to ease monetary conditions.

1964   Title VII of Civil Rights Act reduced degree to which women and men are treated differently under law.

1965 to 1984   Total welfare and unemployment transfer payments (in 1983 dollars) increased from $148.6 billion to $475.6 billion. Despite this more than 200 percent increase, little progress was made in reducing the poverty rate; in fact, for those of working age, the poverty rate actually rose. For those over age 65, however, the poverty rate fell from 22.4 percent in 1965 to 4.3 percent in 1984.

Mid-1960s   Justice Department challenged the anticompetitive nature of fixed commission rates and urged the SEC either to justify or to eliminate them.

1968   By this time, largely because the average number of shares traded by individuals had almost doubled, the "regulatory tax" had increased to $68 million. Commissions paid by institutional investors had risen to $495 million, with $327 million representing charges above competitive rates. The New York Stock Exchange implemented a new structure of fixed commissions that would cut excess payments roughly in half.

1968   Strassburg moved to get control over AT&T; initiated "Computer Inquiry" opening way to independent supply of terminal equipment (telephones, answering machines, modems, etc.). FCC considered application from MCI to furnish service.

1968 to 1978   National defense share of GNP plunged from 9.0 to 4.8 percent.

1969   MCI's application approved by FCC.

1969 to 1971   Finance policies of Great Society and Vietnam War increased inflation, drained deposits from banks. Rises in market interest rates caused by government borrowing.

1970 to 1983   Military work generated returns on assets of 20.5 percent—about 54 percent higher than the 13.3 percent earned on investments in all U.S. durable goods manufacturing.

1970s   American crop values skyrocketed during inflation; farmers entered bidding frenzy for land.

1970s and 1980s   Financial crises revived debate about bank regulation.

1970s, early   Many observers pointed to increasing famine as evidence that the world was running out of resources.

1970s   Lowell again in a slump: Domestic electronics and shoe industries had not kept up with foreign competition, factories were forced to close again.

1970s Technological changes abroad, and expensive labor at home, made it unprofitable to produce steel in the United States. The demise of the steel industry forced Pittsburgh to create a new way of life; many displaced workers ended up with better jobs in high-tech industries.

1970s to 1982 Money market fund assets grew from nothing to $230 billion.

1971 FCC issued its *Specialized Common Carriers* decision, allowing independent microwave facilities to be constructed under very general circumstances.

1971 to 1974 Nixon administration, inheriting difficult political choice, opted for mandatory wage–price–rent controls. Monetary policy became more expansionary after controls were introduced.

1972 to 1974 Ethiopian famine; 200,000 died.

1974 MCI offered "Execunet," utilizing interconnection with the Bell System to offer switched long-distance service to subscribing customers. The FCC, urged on by AT&T, objected to MCI's new service, and the controversy extended into the courts. AT&T was accused of violating the Sherman Antitrust laws; the agreement reached to settle the government's suit would eventually break up the phone company.
Wave of price increases after Nixon's controls ended.

May 1, 1975 SEC permitted its members to set their own prices (commission rates) for the first time in history.

1970s, late Though times were good for farmers, price supports remained in effect; consumers paid over $5 billion more a year for agricultural goods than they would have paid without government programs.
32.5 percent of the U.S. population was in the 20–39 age bracket; 46.3 per cent of the immigrant cohort was in that prime bracket at that time.

1977 Federal Appeals Court agreed with last of MCI's arguments; FCC could not bar MCI from offering service without full evidenciary hearings.

1979 to 1984 Federal outlays on farm programs increased by nearly 600 percent.

1980 As inflation began to level off, American crop prices began to fall.
Many thrifts were on the verge of bankruptcy. Depository Institutions Deregulation and Monetary Control Act allowed thrifts to make a wider range of loans, including adjustable rate loans, and to offer checking accounts paying interest.
Sixteen percent of employed natives were professional and technical workers. The corresponding figure for recent immigrants was 26 percent.
United States had 88 percent of the world's oil wells, but only 14 percent of the world's oil.

1980s, early Disaffection with regulation led to banking deregulation.
Pittsburgh became a model for other former steel towns; representatives from outmoded factory districts as far away as Germany and France came to Pittsburgh to study the way the city had successfully made the switch.

1980s   As in 1890s, military and social spending escalated sharply; also as in 1890s, growing expenditures caused deficit crises and focused national attention on federal taxes.

Historic swing in the balance of payments. Honda and Toyota built plants on American soil.

Lowell, Massachusetts, rebounded in response to the growth in services and high-tech industries.

Ronald Reagan's and Margaret Thatcher's advisers seemed to believe that low savings rates were the prime cause of productivity slowdown, and that overgenerous welfare programs accounted for both.

1981   Government's antitrust suit against AT&T came to trial; in order to keep its vertical integration, AT&T was forced to cede ownership of Bell Operating Companies.

Ronald Reagan elected president with a mandate to "get government off the backs of the people."

1981 to 1984   Number of Federal civilian employees per 10,000 of population increased for first time since 1970.

1981 to 1985   Department of Defense saw an increase of over 40 percent in budget authority.

1981 to 1987   American farms failed at a rate of 3 to 4.5 per thousand, highest rate since Great Depression.

1981 to 1988   Rate of growth of government expenditures very nearly doubled that of the GNP.

1982   Depository Institutions Act extension allowed all depository institutions to offer special accounts with no interest ceilings.

1982 to 1983   Defense firms generated returns on assets exceeding 25 percent.

1982 to 1990   U.S. economy grew without the interruption of recession, one of the longest sustained economic expansions in its history.

1982   ERA defeated when it fell three states short of ratification.

1985   Over $9 billion was transferred to farmers, at a cost of $19 billion. The gain per farmer was $4,000 and the cost per nonfarm laborer was $190.

1986   Federal farm subsidies exceeded $25 billion, representing fully one-third of net farm income.

Immigration Reform and Control Act passed by Congress.

National defense share of GNP reached post-Vietnam high of 6.6 percent.

Tax Reform Act, most significant piece of tax legislation since World War II. United States again largest debtor nation in history; increased demand for tariffs. U.S. household savings ratio was 3.9 percent, compared to 12.7 percent in Germany, 17.0 percent in Japan, and 11.3 percent in Canada.

December 1986   After this time the commission rate per securities transaction reflected the effects of both share price and number of shares.

1986–1991   Defense spending stabilized; rather than cut programs, defense managers pared production runs, resulting in accelerating unit costs and diminished efficiency.

1987   Federal share of GNP was between 20 and 25 percent. Most households had large amounts of debt—from mortgages to credit cards, student loans to credit lines. Private investment was only half as great as the federal budget. Total government expenditures distributed 45 percent of GNP.

October 1987   Stock market lost roughly a third of its value. Alan Greenspan, chairman of the Federal Reserve Board, quickly stepped in and increased the money supply, providing order and liquidity to what might have become a very disorderly market.

1988   Approximately 53 percent of employed blacks held jobs in skilled and white-collar occupations. Fully one-third of all black families enjoyed the social and economic benefits of an income over $35,000.

1990   Immigration Act increased the flow of immigration, set up oversight committee to observe bill's effects.

1990s, early   Rise in coal mining accident rates.

1991   Costs associated with Cold War accrued at about 5 percent.
The minimum margin rate regulated at 50 percent; the $10 stock required an investment of $5 and a $5 loan.
World Bank/IMF considered monetary reform that might place renewed emphasis on the presumedly stabilizing influence of gold.

2000   Twenty of the largest twenty-five cities will be in the Third World.

For Further Reading

I: INTERNATIONAL RELATIONS AND FOREIGN AFFAIRS

Bean, Richard. "War and the Birth of the Nation State." *Journal of Economic History* 33. March 1973. 203–21.

McNeill, William H. *The Pursuit of Power*. Chicago: University of Chicago Press. 1982.

Rosenberg, Nathan, and L. E. Birdzell, Jr. *How the West Grew Rich*. New York: Basic Books. 1986.

II: WORKERS AND EMPLOYMENT

Alston, Lee J. "Farm Foreclosures in the United States During the Interwar Period." *Journal of Economic History* 43. December 1983. 885–903.

Chelius, James R. "Liability for Industrial Accidents: A Comparison of Negligence and Strict Liability Systems. *Journal of Legal Studies* 5. 1976. 293–310.

Fishback, Price V. "Workplace Safety During the Progressive Era: Fatal Accidents in Bituminous Coal Mining, 1912–1923." *Exploration in Economic History* 23. 1986. 269–98.

Gardner, Bruce. *The Governing of Agriculture*. Lawrence, KS: University Press of Kansas. 1981.

Graebner, William. *Coal-Mining Safety in the Progressive Period*. Lexington, KY: University of Kentucky Press. 1976.

Kilman, Scott, and Jean Marie Brown. "Blighted Bounty." *The Wall Street Journal*. November 9, 1987. 1, 21.

III: WOMEN AND MINORITIES

Lacayo, Richard. "Between Two Worlds." *Time*. March 13, 1989. 58–68.

Levin, Michael. *Feminism and Philosophy*. New York: Transaction Books. 1984.

Margo, Robert A. *Race and Schooling in the South 1880–1950: An Economic History*. Chicago: University of Chicago Press. 1990.

Sowell, Thomas. *Ethnic America*. New York: Basic Books. 1981.
Williams, Walter. *The State Against Blacks*. New York: McGraw-Hill. 1982.

IV: GOVERNMENT AND THE ECONOMY

Buckley, Peter, and Brian Roberts. *European Direct Investment in the U.S.A. Before World War I*. New York: Macmillan. 1982.
Danielian, Ronald L., and Stephen E. Thomson. *The Forgotten Deficit: America's Addiction to Foreign Capital*. Boulder, CO: Westview. 1987.
Ferguson, James. E. *The Power of the Purse: A History of American Public Finance, 1776–1790*. Chapel Hill: University of North Carolina Press. 1961.
Krugman, Paul. *The Age of Diminished Expectations*. Cambridge, MA: MIT Press. 1990.
Shepherd, James F., and Gary M. Walton. *Shipping, Maritime Trade, and the Economic Development of Colonial North America*. London: Cambridge University Press. 1972.
Williamson, Jeffrey G. *American Growth and the Balance of Payments*. Chapel Hill: University of North Carolina Press. 1964.

V: REGULATION, DEREGULATION, AND REREGULATION

American Petroleum Institute. *Facts and Figures*. New York: 1951.
Butler, E. F. *Forty Centuries of Price Controls: How Not to Fight Inflation*. Washington, D.C.: The Heritage Foundation. 1979.
Federal Oil Conservation Board. *Complete Record of Public Hearings*. Washington, D.C.: 1926.
Galbraith, John Kenneth. *A Theory of Price Control*. Cambridge, MA: Harvard University Press. 1952.
Libecap, Gary D. "The Political Allocation of Mineral Rights: A Reevaluation of Teapot Dome." *Journal of Economic History* 14. June 1984. 381–93.
Mitchell, H. "The Edict of Diocletian: A Study of Price Fixing in the Roman Empire." *Canadian Journal of Economics*. February 1946. 1–12.
Ragland, Reginald. *A History of the Naval Petroleum Reserves and of the Development of the Present National Policy Respecting Them*. Washington, D.C.: 1944.
Rockoff, Hugh. *Drastic Measures: A History of Wage and Price Controls in the United States*. London: Cambridge University Press. 1984.
Wiggins, Steven N. "Contractual Responses to the Common Pool: Prorationing of Crude Oil Production." *American Economic Review* 74. March 1985. 87–98.

VI: TECHNOLOGY AND COMPETITIVENESS

Cox, Archibald. *The Court and the Constitution*. Boston: Harvard University Press. 1987.

Fuchs, Victor R., ed. *Production and Productivity*. New York: Columbia University Press. 1969.

Griliches, Zvi. "Hybrid Corn and the Economics of Innovation." *Science*. July 29, 1960.

Haites, Eric, James Mak, and Gary M. Walton. *Western River Transportation*. Baltimore: Johns Hopkins University Press. 1975.

Hazlett, Thomas W. "Duopolisitic Competition in CATV: Implications for Public Policy." *Yale Journal on Regulation* 7, 1. January 1940.

Hunter, Louis C. *Steamboats on the Western Rivers*. Cambridge, MA: Harvard University Press. 1949.

Jewkes, John, et al. *The Sources of Innovation*. New York: Norton. 1969.

Kranzenberg, Melvin, et al. *Technology and Culture*. New York: Shocken. 1972.

Mayr, Otto, and Robert C. Post. *Yankee Enterprise*. Washington, D.C.: Smithsonian Institution Press. 1981.

McCloskey, Donald N., ed. *Essays on a Mature Economy: Britain After 1840*. London: Methuen & Co. 1971.

Rostow, Walter W. *Politics and Stages of Growth*. London: Cambridge University Press. 1971.

Shannon, Fred A. *The Farmer's Last Frontier: Agriculture 1860–1897*. New York: Harper. 1968.

Thomson, David. "The Great Steamboat Monopolies, Part I: The Mississippi." *The American Neptune* 16. 1956. 28–40.

Index